Andréa Esteves Martins

Ethnopharmacology in the Denis Gonçalves settlement in Goianá, Minas Gera

Andréa Esteves Martins

Ethnopharmacology in the Denis Gonçalves settlement in Goianá, Minas Gera

Ethnopharmacology in a Settlement

ScienciaScripts

Imprint

Cover image: www.ingimage.com

This book is a translation from the original published under ISBN 978-613-9-73171-8.

Publisher:
Sciencia Scripts
is a trademark of
Dodo Books Indian Ocean Ltd. and OmniScriptum S.R.L publishing group

120 High Road, East Finchley, London, N2 9ED, United Kingdom
Str. Armeneasca 28/1, office 1, Chisinau MD-2012, Republic of Moldova, Europe
Printed at: see last page
ISBN: 978-620-7-88553-4

I dedicate this work to my grandmother Maria Rezende (Sinhá), who taught me the value of knowledge, and to my son Bruno, who has been my great companion on this journey, supporting me and making me realise that life is worth living.

"When we cherish the desire to learn more, our lives will be filled with genuine vitality and (rilho". (Daisaku Ikeda).

ACKNOWLEDGEMENTS

I would especially like to thank my parents for all the opportunities, my grandmother Maria Rezende (Sinhá) for her support and encouragement, my son Bruno, a great companion for his support and encouragement and for showing me the way, the members of the Health Collective of the Landless Rural Workers' Movement: Lenira Nunes da Silva, Milton Rodrigues dos Santos, Delvanir Alves Ribeiro, Maria Lúcia Antonelli Marcelo, Ana Márcia de Almeida Souza, Diliege Vitório de Araújo Lopes, José Soares da Silva, Luiza Margarida Cândida Antonelli, João Ferreira de Souza, Emilse Aparecida Ribeiro, Ana Paula de Jesus dos Santos, Sebastião Antonelli and the former residents in the area (Colonos): Mr Mateus de Souza, Mr Sebastião Marcelino, Mr José Mateus da Silva, Mr José Lourenço Santos Filho, Mr Anézio Batista Lausino and the following ladies: Maria do Carmo de Almeida, Celma Conceição, Maria de Fátima Nascimento Rezende, Maria Aparecida da Costa, Rosilda Célia da Silva Lage, Marina Clara Ferreira Flauzino, Jaucila Almeida Dolatezi, Iranete Luzia Alves de Souza, Leila Regina Alves for their welcome, trust and collaboration, my husband Sérgio for his encouragement, my sisters for their encouragement. To my friend Selma for her encouragement, to my friends, and especially to Amanda Surerus Fonseca for being a great collaborator and companion, always responsible, committed and a partner in the development of the work and to my work colleagues for their encouragement, to my supervisor Daniel Sales Pimenta, to the members of the panel, to the Postgraduate Programme in Ecology at the Federal University of Juiz de Fora. I would like to take this opportunity to express my gratitude, especially to José Carlos, who is no longer with us, for his attention and availability, and also to Júlio and Rose, who were always attentive, ready to answer and clarify all questions relating to the programme, and also to Flávia, who was always helpful and attentive. And to all those who were present during this process.

SUMMARY

Among the main Brazilian biomes is the Atlantic Rainforest, where the Denis Gonçalves Settlement is located, an area of the former Fazenda Fortaleza de Santana, in Goianá, in the state of Minas Gerais. The aim of this study was therefore to carry out an ethnopharmacological study in the Denis Gonçalves Settlement with MST rural settlers and settlers, to assess the transfer of ethnopharmacological knowledge and the degree of vulnerability regarding the use of the main native species used. To this end, the following methods were used: semi-structured interviews; collection of botanical material and obtaining a voucher from the Herbarium; quantitative analysis using the calculation of risk of use and value of use. Comparisons were also made of ethnopharmacological knowledge in relation to scientific literature and the recommendations of RDC No. 10 - ANVISA, to suggest priority implementation in medicinal gardens. Using statistical calculations, the medicinal plants were selected according to their relevance of use for the interviewees and, for those species native to the Atlantic Forest, they were assessed in terms of their current state of conservation in the study area, in order to provide support for the construction of a sustainable management plan. With regard to local ethnopharmacological knowledge, it was possible to inventory 105 species from 41 botanical families in the MST group and 75 species from 32 botanical families in the Colonos group. There was no specific extraction of native species in relation to local medicinal use. The species used in both groups meet local health demands. Therefore, 20 species were listed with the backing of RDC No. 10 - ANVISA, for later implementation in a community medicinal garden, which is currently being set up.

Key words: conservation; phytotherapy; medicinal plants.

SUMMARY

CHAPTER 1	**5**
CHAPTER 2	**6**
CHAPTER 3	**13**
CHAPTER 4	**14**
CHAPTER 5	**15**
CHAPTER 6	**23**
CHAPTER 7	**51**

CHAPTER 1

INTRODUCTION

Tropical forests contain more than half of the world's existing species (KURTZ & ARAÚJO, 2000) and have historically been occupied by human groups who live and interact with them (TICKTIN **et al.** 2012). These forests provide raw materials and form an important part of certain people's lives (VARGHESE **et al.** 2015), including rural communities, such as the Rural Settlers of the Landless Workers' Movement (MST), who are a group occupying delimited spaces intended for agricultural activities, separated from each other and set up by the National Institute for Colonisation and Agrarian Reform (INCRA).

This type of population has accumulated a vast amount of knowledge about the natural resources and respective ecosystems on which they depend for the maintenance of their lives (TOLEDO, 2002; PINTO **et al.** 2006; TOLEDO & BARRERA-BASSOLS, 2008), as is the case with the use of flora for medicinal purposes (VEIGA-JÚNIOR, 2008). This knowledge about the use of medicinal plants is often the only way to treat illnesses and maintain health (VEIGA-JUNIOR **et al.** 2008).

The study of knowledge about the use of medicinal plants is based on ethnopharmacology, which in the context of ethnobotany is interpreted by Albuquerque & Hanazaki (2006) as the study of traditional compounds, including plants, used in health systems. In this way, this knowledge could also provide information on use and management, serving as a subsidy for implementing future initiatives aimed at using natural resources in a sustainable way (CONDE, 2016), since the remnants of this forest in Minas Gerais have great biological richness and are concentrated in isolated fragments, subject to various environmental or economic pressures (DRUMMOND **et al.** 2005).

CHAPTER 2

LITERATURE REVIEW

2.1 MAN AND NATURE CONSERVATION

Anthropogenic and natural disturbances of different durations, intensities and frequencies are considered to be part of nature's history, influencing its current dynamics (BALÉE & ERICKSON, 2006).

The relationship between man and biodiversity is very complex and over time it has alternated between dominating and protecting nature. Furthermore, there are different views on this relationship according to different cultures (AMOROZO, 2007).

The loss of ecological and cultural diversity, poverty and inequality tend to increase the vulnerability of human life and planetary ecosystems (RATTNER, 2009). In Brazil, the direct exploitation of natural resources and the elimination of vegetation cover have caused a rapid loss of natural wealth, which represents a threat to Brazilian ecosystems, and in particular the Atlantic Forest biome, one of the most threatened regions on the planet (CÂMARA, 2002).

2.2 THE ATLANTIC FOREST

The Atlantic Rainforest is one of the most important tropical forests in the world and is also considered to have the greatest biodiversity (CONDE, 2016). It is located in the tropics, on the plains along the coast and mountain escarpments (PAVAN- FRUEHALF, 2000), and is the oldest forest formation in Brazil (LEITÃO FILHO, 1987). At the time of its discovery, it expanded along the eastern Brazilian coast from Rio Grande do Norte to Rio Grande do Sul, occupying an area of approximately 1 million km^2 (PEIXOTO et al. 2002) and which, after 500 years of appropriation, has been reduced to approximately 5% of its original area (AB' SABER, 2003). It is currently found as small patches of forest, with few extensive areas and various stages of degradation (GUATURA et al. 1996), mainly in the south and southeast (MYERS et al. 2000).

2.3 LIVING POPULATIONS IN THE CONTEXT OF THE ATLANTIC FOREST

Brazil's high rates of biodiversity and endemism combined with great cultural miscegenation have resulted in a vast knowledge of flora (RODRIGUES & CARLINI, 2003). However, one can't help but take into account that a large part of the still-preserved areas of national territory are inhabited by traditional communities. In the context of the Atlantic Rainforest, we can observe the occurrence of communities such as indigenous peoples, quilombolas, riverine communities and rural populations (CUNHA & ALMEIDA, 2001), such as rural settlers. The preservation of this biome's environment is therefore fundamental for the maintenance of rural populations, since environmental degradation is verified by various aspects, such as the alarming level of pesticides in the soil and groundwater, mass extinction of species, among others (CONDE & PIMENTA, 2015).

2.4 MAN'S RELATIONSHIP WITH PLANTS

Throughout his history on the planet, man has accumulated a wealth of information about natural resources and has also learnt to use these resources for his survival needs (AMOROZO, 1996), which has been the case mainly with plants, which were and still are used for various purposes such as food, building houses, making various utensils, as medicines, among other uses (DIEGUES, 2000), and which is relevant in Brazil because it has one of the richest floras in the world, 99% of which are chemically unknown (GOTTLIEB **et al**. 1996).

Since the emergence of ancient man, who lived in tropical forests for thousands of years, they had already become dependent on plants (SZABÓ, 1996).

The history of the use of therapeutic plants is ancient: the Chinese already used flora 3000 BC and to this day China's official medicine is herbal. In the same way, the history of experimental pharmaceuticals begins with Mitriades, king of Porto in the 2nd century BC, who is considered to be the first experimental pharmacologist. Opiates and numerous toxic plants were also known at this time (ALEIXO, 1995). In the Ebers papyrus from 1550 BC, discovered in the middle of the 20th century in Luxor, Egypt, around 700 drugs were mentioned, including plant extracts, metals and animal venoms from various sources (ALMEIDA, 1993). The Bible also contains many references to healing plants such as aloe, benzoin and myrrh. In ancient times, in Greece and Rome, medicine was also related to botany, as in the work of Hippocrates, Corpus Hippocraticum, indicating a plant-based remedy for each illness (MARTINS **et al.** 2000).

In Brazil, the treatment of diseases with the use of plants is influenced by various cultures such as indigenous, European, African and Asian (ALMEIDA, 2000).

2.5 ETHNOBOTANY AND ETHNOPHARMACOLOGY

Ethnobotany is the discipline concerned with studying the direct interrelationship between people and plants (ALBUQUERQUE, 2005) and is defined by Heinrich **et al.** (2004) as the study of the relationship between human populations and plants in all their complexity.

One of the branches of ethnobotany is the study of medicinal plants. In this case we have ethnopharmacology, which according to Albuquerque & Hanazaki (2006) is a discipline that deals with the study of traditional preparations used as therapeutic resources. Ethnopharmacology is at the intersection of medical ethnography and the biology of therapeutic action, i.e. a transdisciplinary exploration that encompasses the biological and social sciences (ETKIN & ELISABETSKY, 2005), and one of its objectives is to value traditional knowledge. According to Amorozo & Gely (1998), it can also serve as an important tool for maintaining traditional life systems, minimising the loss of a complex body of knowledge and a genetic heritage of inestimable value for future generations (AMOROZO & GÉLY, 1988), and in relation to traditional life systems there are the rural populations made up of settlers from the Landless Rural Workers' Movement and settlers, who occupy the area of the former **Fazenda Fortaleza de Sant'Ana, located in a fragment of Atlantic Forest.**

2.6 THE SANTANA FORTRESS FARM

2.6.1 Historical aspects

Its history began at the end of the 18th century with the Pereira de Souza family's right of possession, acquired by a sesmaria letter issued in 1811 (COLOMBO, 2007).

In 1815 this property was called Fazenda Fortaleza do Rio Novo, and the name fortress was given to the farm because of the rocky formation of the Serra da Babilônia. And then Fortaleza de Santana due to the mixture of the past name, Fortaleza, with **Sant'Anna for being the patron saint of the community (COLOMBO & CORRÊA, 2014).**

In the mid-19th century, the property was sold to Captain Mariano José Ferreira Armond, who moved into the area with his wife Maria José de **Sant'Anna and their son Mariano Procópio Ferreira Lage, who later invested in** coffee **production**, turning Fazenda **Fortaleza de Sant'Anna into one of the largest** coffee **producers of** the 19th century (GUIMARÃES, 2009).

The farm was visited by the Swiss naturalist Louis Agassiz, King Pedro II, the French diplomat Conde de Gobineau, Getúlio Vargas, Juscelino Kubitschek, Antonio Carlos Andrada, Benedito Valadares and many others (COLOMBO, 2007).

In 1871, three caves were discovered in the Serra da Babilônia with large quantities of skeletons, mummies, funerary urns and various objects. All this material was donated to Emperor Pedro II, who sent it to the National Museum in Rio de Janeiro (COLOMBO & CORRÊA, 2014).

At the end of the 19th century, the farm was in crisis due to the fact that Mariano Procópio had left it debts. However, in 1872, with the death of Mariano Procópio, the property passed to Frederico Ferreira Lage, who invested in technology and hired Italian and German families, even before abolition. All the investment didn't work out, the farm was mortgaged and later auctioned and bought in 1902 by Cândido Ferreira Tostes, who invested in coffee production and re-established the system of colonies, mainly of **Italians, which gave him the title of "The Coffee King" (SARAIVA 2005).**

In March 2001, the Fazenda's headquarters were completely destroyed by a fire, losing a priceless historical collection such as antique furniture, rare books, works of art, old photos, diplomas and medals (COLOMBO, 2002).Even with the fire and the loss of the headquarters, the Fazenda continues to be a demonstration of diverse historical and cultural experiences, making this property recognised by the public authorities for the protection of cultural heritage (Figure 1) (SILVA, 2007).

Figure 1: Area of the Casa Grande and other buildings on the former Fazenda Fortaleza de Santana and the current Denis Gonçalves Settlement. Image credit: Inter- relações arquitetônicas. Available at: http//www.ihgrgs.org.br. Accessed in April 2014.

2.6.2 Cultural Aspects

The Fazenda Fortaleza de Sant'Anna has cultural heritage with potential for archaeological research, with old roads, retaining walls, cemeteries, several mummies found in the Serra da Babilônia, ruins of buildings that are hydraulic dams built at high points in the Serra da Babilônia or by diverting streams, as well as constructions linked to agricultural production, such as tulhas and coffee processing houses, with elements of Germanic and Italian architecture such as the "Vila dos Colonos" (Settlers' Village), **documents and works of art such as the image of Sanfanna, the community's patron saint** (COLOMBO, 2007).

At the time of colonial Brazil, there were forms of labour relations that were established between landowners and rural workers, known as settlers, but these workers could also be subsistence tenants or wage earners (FALEIROS, 2007).

In the transition from slave labour to wage labour, immigrant families arrived from Europe, from countries such as Italy and Germany. Together with the former slaves, they made up the group of settlers living on the former Fazenda Fortaleza de SanfAnna (COLOMBO & CORREA, 2014).

2.6.3 Environmental characterisation of the area studied

According to Taveira (2010), in the area of Fazenda Fortaleza de Santana there is Semideciduous Seasonal Forest, conditioned by climatic seasonality, with a dry period and a rainy period, with deciduous trees, average temperatures of 21°C, original vegetation replaced by pastures with grasses, bathed by the waters of the Santana, Laje, Goianá and Ponte Preta streams, tributaries of the Novo river and the Retiro stream, with a relief featuring hills and lowlands.

2.7 RURAL SETTLEMENTS

According to Ribeiro (2004), rural settlements are human groups in which studies in the field of ethnopharmacology are promising, since they live integrated into the natural environment and draw their main resources from local nature.

These are socio-territorial movements whose aim is to conquer land for work by occupying unproductive estates. Fernandes (2000) considers them to be local social units with identities based on the experience of common experiments (NEVES, 1999), spaces of civility that spread across the countryside (D'AQUINO 1998), where the collective history of the rural population is explained and modified (CAPPELLIN & CASTRO 1997).

2.7.1 The Landless Workers' Movement (MST)

The MST originated in 1984 in Cascavel, PR (BRENNEISEN, 2013) when hundreds of rural workers decided to set up an autonomous peasant social movement that would fight for land, agrarian reform and important social changes for the country. The movement was formed by squatters, people affected by dams, migrants, sharecroppers, partners and small farmers who were landless rural workers (MST, 2016).

This struggle has its origins in historical and cultural factors in the formation of Brazilian society, and the possibility of land distribution and occupation became an eminently political responsibility (TEIXEIRA, 2012), which led to the construction of a new constitution, approved in 1988, with the achievement of articles 184 and 186, which guaranteed the expropriation of land that did not fulfil its social function (MST, 2016).

According to Comparato (2001), the MST is a different form of political action, based on methods and principles that are already known, and on the political competence that the movement has demonstrated by making allies in various sectors of civil society. The MST also organised the First Women's Meeting, which created an initial space to debate the gender issue, as well as women's militancy and participation in political decision-making in relation to the movement, and it was also discussed that women should be declared as settlers (FARIAS, 2016).

2.7.2 The Denis Gonçalves Settlement: History and Trajectory

The Denis Gonçalves Settlement came about with the occupation by 50 families of the Fortaleza de Santana farm on 25 March 2010 at 5.50am, which was also attended by MST activists, students from the Federal Universities of Juiz de Fora and Viçosa, trade unions from Juiz de Fora and the Pastoral Land Commission (CPT).

According to ALMEIDA (2014), in 2011, 90 families were evicted from the area of Fazenda Fortaleza de Santana, and then occupied the roadside near kilometre 48 of the MG 353, remaining in this condition while awaiting a court injunction to reoccupy the farm.

According to the settlers themselves, the occupation of the farm was a long and painful process that began with the arrival of the families on the farm, which they occupied for a year and were then evicted by a court order. The eviction took place with the participation of the police, who arrived on horseback, using helicopters and in addition to evicting them, they also destroyed their crops. They then began to occupy the roadside near kilometre 48 of the MG 353 and say that it was a difficult time with a lot of struggle, without a minimum of dignified living conditions, without any structure, living in tents made of tarpaulin, on the side of the road, thus being exposed to the cold, heat, abundant rain, lack of basic sanitation, electricity, medical assistance and also various dangers.

According to TEIXEIRA (2012), in July 2010 there was the first hearing at the Rio Novo Municipal Forum, but without an agreement, the hearing continued without results and the workers had to return to the camp to wait for the legal proceedings to be completed. It was then, in December 2011, that a decree was issued authorising INCRA to begin the legal procedures for expropriating the farm. The decree was published in the Federal Official Gazette, and so they were finally able to return to the farm for the occupation process with the new name in honour of the member of the Olga Benário Settlement, who died in an accident and was renamed the Denis Gonçalves Settlement.

2.7.3 Health Issues in the Denis Gonçalves Settlement

The MST's Health Sector prioritises the battle for health as a human right, participating in the development of a popular project for the working class, built on the concept of plenitude and equality, reasoning about the causes of individual and collective illnesses, seeking a new meaning for health based on an ethical, political, social, economic and cultural dimension, rescuing the value of the farmer's knowledge, of popular knowledge (FARIAS, 2016).

For this reason, the Health Sector has the role of pushing the state to fulfil its role in settlements and encampments, and to implement public policies for sovereignty, food security and decent living conditions, with preventive attitudes towards diseases (MST, 2016).

And with regard to health protection measures, it is clear that learning about alternative preventive practices, healing with the use of plants and teas, have become part of the community's health maintenance objectives, with workshops on the use of herbal medicines and other alternative therapies, giving health greater sustainability with regard to natural resources, by aiming at agrarian and productive restructuring, in the sense of social justice and agroecology (FARIAS , 2016).

And if we're talking about natural resources, in the historical context, women also had knowledge of plants with healing powers, which they received orally from their mothers. Women who came from regions such as Zona da Mata and Vale do Rio Doce, with their differentiated knowledge, went on to form the MST's Health Group, called the "Health Collective" (FARIAS, 2016).

CHAPTER 3

OBJECTIVES

3.1 MAIN OBJECTIVE

Carrying out an ethnopharmacological study among the groups of rural settlers of the MST Denis Gonçalves and Colonos, living in the area of Fazenda Fortaleza de Santana, in the municipality of Goianá - MG.

3.2 SPECIFIC OBJECTIVES

To assess the maintenance of traditional knowledge about medicinal plants among the groups evaluated: MST and Colonos;

Recording local knowledge about medicinal plants;

To assess the vulnerability of medicinal species, based on the frequency, value and risk of their use in the MST and Colonos groups;

Correlating epidemiological data with the medicinal plants used and their preferred use, with a view to sustainability in health in the MST and Colonos groups.

CHAPTER 4

BACKGROUND

Studies into the use that communities make of natural resources can be an important subsidy for drawing up projects that also aim to maintain local cultural knowledge. By combining the knowledge of natural resources held by traditional populations with scientific knowledge, strategies can be developed to make better use of this knowledge, increasing the possibilities for development in these areas (ALBUQUERQUE & ANDRADE, 2002). The wealth of knowledge about natural resources held by traditional populations also provides science with models for the sustainable use of natural resources Silva **et al** (2015) and according to Brito and Valle (2011) the knowledge of traditional populations supported by scientific knowledge corroborates the coherent use of medicinal plants to raise awareness about the importance of conserving them.

This study is justified in the sense that it is aimed at recording ethnopharmacological knowledge and assessing anthropogenic pressure on the natural resources of the flora in question, thus serving as a strategy to subsidise future initiatives, such as the production of booklets and management plans aimed at conserving biodiversity and local culture. In this way, ethnopharmacology presents itself as a tool for bringing together the groups living in the area in question (settlers and MST), since it will seek to record and perpetuate the collection of local knowledge in the area studied with a collective character that can provide interactivity between them.

CHAPTER 5

MATERIALS AND METHODS

5.1 STUDY AREA

The research was carried out in the Denis Gonçalves Settlement, in the municipality of Goianá, in the state of Minas Gerais (Figure 2).

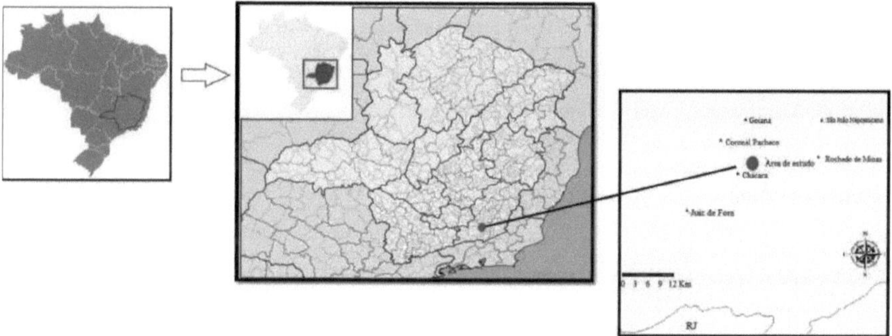

Figure 2: Location of the Denis Gonçalves Settlement in Minas Gerais, Brazil and nearby towns (Juiz de Fora; Chácara; Coronel Pacheco; Goianá; São João Nepomuceno; Rochedo de Minas.

5.2 SAMPLING: PEOPLE INVOLVED IN THE STUDY

The groups involved in the ethnopharmacological study were residents of the Denis Gonçalves Settlement in Goianá, Minas Gerais, made up of participants from the Health Collective of the Landless Rural Workers' Movement, and the Colonos, former residents of the former Fazenda Fortaleza de Santana, now the Denis Gonçalves Settlement.

The MST Health Collective group was made up of 12 medicinal plant experts and recommenders, 10 of whom had already been selected, the other two joined the group during the data collection phase of this study, as well as another medicinal plant expert who was not part of the aforementioned Collective, totalling 13 interviewees.

The Colonos group is represented by 14 connoisseurs and recommenders of medicinal plants, who are descendants of the old Colonos, represented by slaves and German and Italian immigrants.

The interviews were carried out with the Health Collective group, which had previously been chosen by the MST leader, but in the case of the settlers the Snowball technique was used to choose the participants

(COTTON, 1996), where an influential resident among the settlers nominated other settlers who stood out in terms of their knowledge of medicinal plants, who in turn nominated others to be interviewed (MAROYI, 2011).

5.3 PRESENTATION OF THE INTENDED WORK AND OBTAINING A LICENCE FROM THE RESEARCH ETHICS COMMITTEE

To carry out the research, authorisation was first sought from the Ethics Committee of the Federal University of Juiz de Fora (Annex 1). A meeting was then held with the community, together with the leader of the MST, Ludmila, and members of the Denis Gonçalves Settlement Health Collective, where presentations were made about the research. The meeting took place in front of the farm's coffee-processing machinery.

In the case of the settlers, there was a family visit to all the homes where people were willing to take part. In this case, Mrs Iranete Luzia took part and drove the researcher to their homes to carry out the interviews.

According to Albuquerque & Lucena (2004), key informants are people who have influence in the communities and can contribute more actively to the research. In the case of this research, the key informants were MST leader Ludmila and Mrs Iranete Luzia, a resident with great influence among the settlers.

5.4 METHODOLOGICAL TOOLS

5.4.1 Strategies for approaching research participants

In order to achieve better acceptance by the groups with whom the ethnopharmacological research was to be carried out, we began to implement strategies to bring the researchers closer to the groups and their leaders, with meetings for introductions.

During the course of the work, other meetings took place, such as the joint effort to start building the Denis Gonçalves Settlement health centre. This event was attended by members of the MST group, settlers and other groups that were also working in the settlement, bringing the groups closer together.

In the case of the settlers, the interviews and meetings took place in their homes, accompanied by their leader Iranete Luzia.

There was a fundamental difference between the interviewees from the MST and the settlers. The MST interviewed qualified representatives of the community who were knowledgeable about medicinal plants, while the Colonos interviewed all the families in a quantitative rather than qualitative way. This is due to the characteristics of the two communities: the MST showed organisation and made its members responsible (as in the case of the "Health Collective"), while the Colonos showed no organisation, were disorganised and unmotivated, with only Iranete Luzia standing out among them as an organiser.

Those involved in the research were receptive and agreed to take part in both the interviews and the guided tours, which is a method of working in the field, with walks with guides or plant connoisseurs, to identify the vernacular names of plants and animals within the forests with their attributes and the ways in which they are used, according to Albuquerque & Lucena (2004). The walks with the group of settlers did not take place inside the forest, but on some paths near the MST camp, in the vegetable gardens, backyards and gardens and in the homes, in the case of the settlers' group.

5.4.2 Ethnopharmacological data collection

The interviews were carried out with a group formed by the Health Collective, democratically elected by vocation and voluntarism by the MST itself. In the case of the settlers, there was no closed group and the interviews were carried out in their homes, directed by the key informant, Mrs Iranete Luzia.

Semi-structured forms were used (ALEXIADES, 1996), in which aspects such as: origin of the interviewee; gender and age group of those who fall ill the most, diseases that most affect the groups, learning about medicinal plants, whether they use other non-plant resources, preferences regarding the use of medicinal plants or industrialised medicines, whether they have access to the public health system, whether they use medicinal plants; their forms of preparation and posology; their main uses; where and how they collect them; who they learned about them from; whether knowledge about medicinal plants is being passed on to their descendants (Annex 2).

Figura 3: Interview; A: Maria do Carmo de Almeida and Mrs Celma Conceição; B: Mrs Delvanir Alves Ribeiro (Mrs Neguinha), C: Mr José Soares da Silva, D: Milton Rodrigues dos Santos, E: Mrs Maria Lúcia Antonelli, F: Mrs Maria do Carmo de Almeida. Image credit: Amanda Surerus Fonseca.

5.4.3 Botanical collection, identification and herbarium deposit

At the end of the questionnaire, the interviewees were asked to take the researcher to the place where the species mentioned occurred using the guided tour method (Figure 4) (PHILLIPS & GENTRY, 1993), where in vivo botanical material was collected and photographed. Exsiccates were then produced and identified by a team of specialists from the UFJF Botany Department. The materials were then deposited in the Padre Leopoldo Krieger Herbarium - CESJ, where the respective Vouchers were obtained. The scientific nomenclature was revised and updated according to: http://www.tropicos.org/.

Figura 4: Botanical collection using the Wolking in the Wood method, with the participation of MST residents José Soares da Silva and Milton Rodrigues dos Santos.

5.4.4 Conservation status of native species

In order to find out the risk level of the medicinal species used, we first selected those that were native to the local biome, followed by searches on the Biodiversity Portal (https://portaldabiodiversidade.icmbio.gov.br/), Flora Brasiliensis http://florabrasiliensis.cria.org.br/) and Plantas da Floresta Atlântica (STEHMANN et al. 2009), to find out which were native to Brazil and the Atlantic Forest biome. Afterwards, their conservation status was checked at: IUCN - Red List of Threatened Species (2014); Ministry of the Environment (2008); Biodiversitas Foundation (DRUMOND, 2009) and CNC - Flora (2016).

5.4.5 Quantitative analyses of the data obtained

To better assess the degree of vulnerability of the species listed, statistical analyses were carried out, taking into account frequency of use; use value (PHILLIPS & GENTRY, 1993) and part use based on the calculation of risk of use (DZEREFOS & WITKOWSKI, 2001).
The frequency of citations of the species surveyed was calculated using the number of times each species was cited by the interviewees and the percentage of each species respectively.

To calculate the frequency, we have:

$$F = Cit\ T / P\ Cit$$

In which:

F = total percentage of possible citations for each species in relation to the sample universe of each group studied (MST and Settlers)

Cit T = total citation of a species per group studied

P Cit = maximum citation possibility for a species in one of the groups studied

To calculate the utilisation risk (RU) for the botanical part collected, a methodology was used in which the degree of risk is assessed based on the utilised part of the species (DZEREFOS & WITKOWSKI, 2001):

$$RU = 0.5(C) + 0.5(U) \times 10$$

Where:

C = refers to the botanical part used and its respective management, where the score based on botanical collection management is based on the following parameter:

Individual and descendants (for any type of species) =10
Perennial structures without death (for any type of species) = 07
Permanent aerial structures (for any type of species) = 04
Transient aerial structures (for species with sexual reproduction and perpetuation by seed) = 01
Transient aerial structures (for species with asexual reproduction and vegetative perpetuation) = 0

U= Highest value between diversity of use (Div) and local use (L)

In which:

Div = corresponds to adding one point for each different medicinal use category, up to a maximum of 10 points.

L = corresponds to citation frequency, where the score follows the following parameter:

High (listed by more than 20% of the population) = 10
Moderately high (10 to 20 %) = 07
Moderately low (< 10% citation) = 04
Low (> 10% citation) = 01

Lucena et al. (2013) state that the Use Value Index (IVU) can be used as a complement, identifying the best known and most used species, helping to diagnose a greater incidence of use. Therefore, the UVI will be used as a complement to the RU.

The following formula was used to calculate the UVI (Phillips and Gentry, 1993); adapted by Rossato et al.(1999):

$$\textbf{IVU = LU/n}$$

Where:

U = Number of categories of use mentioned for the species depicted. **n** = Total number of informants interviewed.

In order to relate the indices proposed for assessing floristic conservation in relation to the pressure of use and management, and for establishing the list of priority species for conservation, the following calculation is proposed:

$$\textbf{Cat = \{(IVU x 10)+ RU\}/2}$$

In which:

Cat = corresponds to the final value of the index ratio, with the submitted species categorised accordingly:

Category 1 (species with a score > 55): There is a risk in terms of how *they are collected and they* should no longer be collected;
Category 2 (species with a score between 35 and 55): Should be collected moderately, they may be included among the vulnerable;
Category 3 (species with a score < 35): They are not at risk in terms of how they are collected.

5.4.6 Categorisation of diseases according to ethnopharmacological knowledge

Diseases were categorised according to the International Classification of Diseases (ICD-10) that affect the community, in order to assess the extent to which ethnopharmacological knowledge can support demands in relation to the health problems presented by local residents.

5.4.7 Comparison of the local use of the species listed with pharmacological evidence based on scientific literature

Surveys were carried out to support the popular use of medicinal species based on scientific literature obtained through reviews of the main databases, such as Chemical Abstracts; International Pharmaceutical Abstracts; Latin American and Caribbean Health Sciences Literature; Analytical

Abstracts; OLDMEDLINE; PUBMED, and ANVISA RDC No. 10, to subsidise safe use.

CHAPTER 6

RESULTS AND DISCUSSIONS

6.1 SAMPLE OF PEOPLE INVOLVED IN THE STUDY

The results will be presented in percentages of a total of 13 interviewees among the selected MST members and 14 interviewees among the settlers, with the characterisation of the groups already presented in the methodology. In terms of occupation, of the 13 interviewees in the Health Collective group, 12 are farmers and one is a housewife. In the case of the Settlers, five are farmers; three are housewives; four are retired farmers; one is a primary school teacher and one is a radiology technician. It can be seen that the majority carry out activities linked to the natural environment, which, according to Conde & Pimenta (2015), leads to the habit of taking the main resources for maintaining their lives from nature itself.

6.2 SOCIO-CULTURAL DATA

With regard to average age, the MST interviewees found that the average was 50.5 and the settlers 58.5. Lima **et al.** (2000), Hanazaki **et al.** (2000) and Pavan-Fruehauf (2000) found similar average ages. It can be seen that the older the person, the greater the accumulation of knowledge about plants, as had already been detected by Voeks & Leony (2004), Cakilcioglua **et al.** (2011), and Silva **et al.** (2012). Galeano (2000) states that results like these may indicate that knowledge is being lost among young people. In addition, according to Torres-Avilez **et al.** (2014), accumulated knowledge can be compromised at advanced ages due to casual memory failures or degenerative diseases. This was the case with a prominent medicinal plant expert among the settlers, who is around 86 years old and a reference throughout the settlement, but who was not interviewed due to memory impairment.

As for gender, in the MST Health Collective, 30.76% of those interviewed were men, while in the Settlers' group 35.7% were men, which according to (SILVA **et al.** 2011) may mean that the accumulation of knowledge about medicinal plants is generally greatly influenced by social factors such as age and gender, which constitute various patterns of knowledge and intra-cultural variations, corroborating Pastore (2005), who previously stated that in rural communities it is common for women to take on the role of caring for the health of their families and because they are also the main connoisseurs of the uses of medicinal plants, because according to Farias (2016) they present the wisdom of healing through plants, the preparation of teas and syrups, they are women who hold knowledge that is not accepted by a large part of society.

As for religion, in the case of the MST, 70% are Catholics, 20% Evangelicals and 10% Spiritists. In the case of the settlers, 100% are Catholics, demonstrating the strong influence of the Catholic religion in Brazil's historical and social process (SANTOS, 2013), where coffee and milk plantation owners imposed the sharing of European culture on their slaves (ALMEIDA, 2000). There was no reference to Afro-Brazilian religions, which have a closer relationship with nature and specifically with medicinal plants (ALMEIDA, 2000).

With regard to place of origin, 100% of the interviewees have a rural background and come from the area itself or from the north, south and centre of the state of Minas Gerais. According to Pinto **et al.** (2006), knowledge about medicinal plants of family origin is directly influenced by the ancestral context and place of origin of the interviewee. There was no difference to be noted in the indications for medicinal plants in this study, despite their different origins. It is possible that the MST and Colonos have particularly similar knowledge, which may be related to the fact that their lives have always been linked to the rural environment and the context of the Atlantic Forest in the state of Minas Gerais.

With regard to knowledge of medicinal plants, in the MST, 90% said they had learnt from their parents, 10% from root growers, courses and books. In the settler group, 100% learnt from family members. The transmission of knowledge from older descendants stands out, which according to Amorozo (1996) may mean that this knowledge was perpetuated orally through intense contact between generations, especially in domestic and kinship groups.

6.3 THE RELATIONSHIP BETWEEN HEALTH, GENDER AND AGE OF THE GROUPS STUDIED

In the MST group, it was possible to detect that the male gender is among those who fall ill the most (70%), and among the age groups, children (0 to 15 years old) stand out. Among the settlers, males (85%) are also the most affected by health problems, and among the age groups, children are also most often affected (35%).

As for gender, Burille & Gerhardt (2014) stated that men are more vulnerable to illnesses, especially serious and chronic illnesses, and that they die earlier than women. However, the results found in terms of age group contradict the findings of Viu et al. (2010), who state that in these types of rural communities, the elderly are more prone to health problems. This data seeks to delve deeper into epidemiological information, which is difficult to obtain in communities traditionally excluded from government attention, and which should be correlated with the medicinal plants indicated in the respective communities.

6.4 PREVALENCE OF DISEASES

Among the most common health problems reported by the MST, diseases of the respiratory system stand out (39.39%), with flu being the most cited (46.15% of the respiratory system category), followed by diseases of the circulatory system (15.15%), with hypertension being the most cited (80% of the circulatory system category) (Table 1).

Table 1: Percentage of citations for each disease category; percentage of the most cited disease in each category, for the **MST** group, following the classification criteria of the International Compendium of Diseases - ICD 10 (http://www.cid10.com.br/).

Categories	% of citations (n=100)	% of citations of the most cited disease within the category
Diseases of the respiratory system	39.39%	46.15% (Flu)
Diseases of the circulatory system	15.15%	80% (Hypertension)
General symptoms and signs	15.15%	60% (Headache)
Diseases of the musculoskeletal system and connective tissue	12.12%	50% (Back pain)
Diseases, symptoms and signs related to the digestive system and abdomen	6.06%	50% (stomach problems)
Other infectious and parasitic diseases	6.06%	50% (Virus)
Diseases of the skin and subcutaneous tissue	6.06%	100% (Allergies)

The most commonly cited illnesses in the group of MST members were influenza and hypertension, while in the case of the settlers they were diabetes followed by influenza and hypertension. According to Gonçalves (2015), health problems predominate among children, which may influence the results, since this younger age group in rural populations is often more vulnerable to problems with the respiratory system. And with regard to problems related to the circulatory system, such as hypertension, in general populations, around 20% of adults are hypertensive, regardless of gender, but there is also a noticeable upward trend with age, which is related to the results regarding high blood pressure problems and the increase in age among those interviewed.

Among the settlers, endocrine diseases stand out (26.09%), with diabetes being the most frequent (83% of the endocrine system category), followed by diseases of the respiratory system (21.74%), with influenza being the most cited (100% of the respiratory system category) and diseases of the circulatory system (21.74%), with hypertension being the most cited (100% of the circulatory system category) (Table 2).

Table 2: Percentage of citations for each disease category; percentage of the most cited disease in each category, for the group of settlers and following the classification criteria of the International Compendium of Diseases - ICD 10 (http://www.cid10.com.br/).

Categories	% of citations (n=100)	% of citations of the most cited disease within the category
Endocrine, nutritional and metabolic diseases	26.09%	83.34% (Diabetes)
Diseases of the respiratory system	21.74%	100% (Flu)
Diseases of the circulatory system	21.74%	100% (Hypertension)
General symptoms and signs	13.04%	33.34% (Headache and sore throat)
Diseases of the musculoskeletal system and connective tissue	8.70%	100% (Back problems)
Diseases, symptoms and signs related to	4.35%	100% (stomach ache)

the digestive system and abdomen		
Diseases of the genitourinary system	4.35%	100% (Kidney problems)

In the case of diabetes, it is a degenerative disease that affects the elderly more (COELI, **et al.** 2002) and not younger people as mentioned in item 6.2 and in the case of the settlers, the prevalence of diabetes may be related to the high degree of consanguinity, since they are direct descendants of the former settlers of the farm. According to Santos (2013), the high prevalence of people with certain common diseases may have hereditary causes, one of the main factors being consanguineous marriages among these populations.

6.5 ORIGIN AND TRANSMISSION OF ETHNOPHARMACOLOGICAL KNOWLEDGE ABOUT MEDICINAL PLANTS

With regard to learning about medicinal plants, all the interviewees from the two groups (MST and Colonos) reported multiple sources of learning about medicinal plants, although most of this knowledge came from parents and grandparents, with the MST (90%) and Colonos (100%) reporting more than one option as a source of learning about the use of medicinal plants (Graph 1).

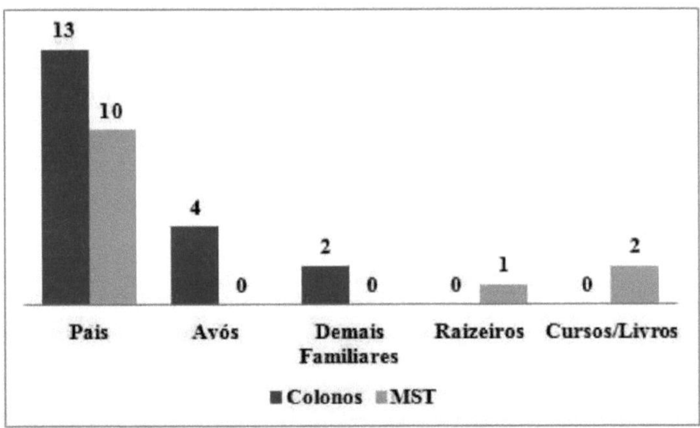

Figure 5: Results in numbers in relation to learning about the use of medicinal plants by the MST and Colonos groups.

According to Toledo & Barrera-Bassols (2010), this type of knowledge is perpetuated mainly orally and between generations of family members. The transfer of knowledge about medicinal plants in rural populations, for the production of medicines, has the family as the main holder of this knowledge, and even today much of the transmission of this knowledge takes place within the family itself, despite some variations in relation to age and gender (MENDIETA et al. 2014).

Regarding the transmission of knowledge about medicinal plants, 92.30 per cent of the MST group reported passing on their knowledge, and 50 per cent of the settlers. This result shows that among the MST group, knowledge is passed on more effectively, while among the Colonos it is 50%.

This may be related to the youth mobilisation initiatives taking place in the group, which may be helping to reduce cultural erosion. However, among the Colonos group, there may be cultural erosion of this type of knowledge. This fact was confirmed in informal conversations with the elders, who said that the current lack of paid work in the area has led to a lack of interest among young people in the area's natural resources. Evasion by younger people has also been found in quilombola communities in ethnopharmacological surveys carried out by the same research group (Siqueira, 2014; Rogério, 2014). STANISKI; **et al.** (2014) further justify this loss of knowledge because there are countless intracultural interferences or variations suffered in modern times, such as migration to work in urban areas and the technological facilities of modern life.

6.6 NON-PLANT THERAPEUTIC RESOURCES

Regarding the use of non-plant natural resources, 38.46% of the MST group use other therapeutic resources (minerals and animals), namely: clay, crystal, clay, pitchstone, duck gizzard, lizard and snail. Among the group of settlers, 7.14%, one person, uses clay and mud. These results show that the MST group has a greater choice of medicinal resources. Helfand & Cowen (1990) cite that there are several records on the use of animal and mineral resources by humans, however, Rodrigues **et al.** (2012), states that in addition to being less used, ethnopharmacological studies that report the use of minerals and animals as therapeutic resources are rare, which ends up making it impossible to compare this cultural aspect in relation to the type of group studied, demonstrating the importance of collecting such data in ethnopharmacological studies.

6.7 PREFERENCE FOR THE USE OF INDUSTRIALISED MEDICINES AND MEDICINAL PLANTS

With regard to industrialised medicines, 60% of the MST group uses them, while 100% of the settlers do. This indicates that although all the residents have access to the public health system, as we have already seen, the MST group still uses natural resources, signalling that, comparatively, this group is promising in terms of self-management and commitment to survival and building alternatives according to their demands, such as, for example, the maintenance of private medicinal and food subsistence gardens, which are commonplace among the MST and rarely seen among the settlers.

Regarding the preference for the best resource, between medicinal plants and industrialised medicines; in the MST group, 90% prefer medicinal plants, and in the Settlers group, 57%. This data shows that among the MST group, the trust and preference for the use of medicinal plants suggests that they still live in a way that is integrated with the natural environment, using the natural resources available. According to Antonio **et al.** (2013), this may indicate that this group maintains respect for traditional knowledge, as well as using more alternatives to promote health, promoting environmental sustainability with local development. On the other hand, among the Colonos, the considerable number of people who prefer to use industrialised medicines indicates that the improved economic condition they have experienced, through income redistribution programmes and policies for distributing medicines in the public health service, as reported informally, added to cultural erosion, may result in low use of cultural knowledge about the use of medicinal plants (PINTO **et al.** 2015). According to the members of the Settlers' group, and based on informal reports, industrialised medicines have become more widely used because they are easier to obtain, but some people have stopped using medicinal plants because they no longer have the physical conditions to plant and maintain vegetable gardens.

The MST group's concern for sustainability in health, with regard to the use of natural resources, is promising, demonstrating the importance of aggregating the community associated with traditional knowledge. According to Farias (2016), improving and supporting people's health is fundamental for successive social development and helps to increase quality of life, and should be geared towards taking advantage of traditional knowledge, with scientific research into flora being a conditioning factor for the gradual improvement of the health system.

6.8 ACCESS TO THE SINGLE PUBLIC HEALTH SYSTEM (SUS)

Regarding access to the public health system, (100%) of the participants in both groups (MST and Colonos) reported that they use it and are satisfied with it, and 100% reported that they use the same service. According to Halal et al. (1994) satisfaction is associated with the resolution of the problem, as well as access to medicines in health centres. However, the prioritised use of ethnopharmacological knowledge among the MST group may suggest that the group's aggregation promotes cultural relevance, which occurs through their self-determination.

6.9 FUTURE PROSPECTS: BUILDING A COMMUNITY MEDICINAL GARDEN

In the Denis Gonçalves settlement, the construction of a vegetable garden with medicinal plants that takes into account local ethnopharmacological knowledge is a verified demand, and one that was already being implemented during the data collection of this research. All the interviewees from the MST and the settlers approved of the idea of building such a garden and would also like to see meetings to exchange knowledge about medicinal plants and the environment.

As observed, the research participants see the vegetable garden as a space for exchanging information, which, according to Conde (2012), qualifies it as a strategy for transmitting environmental education in the pursuit of sustainability, and can promote the maintenance of knowledge about the use of local medicinal plants, as well as serving as an integrating space for strengthening and bringing community capacity closer together, which would serve as an important tool for bringing MST members and settlers closer together. Oliveira & Menine (2012) also point out that a vegetable garden in a given community could bring people together. According to Sacramento (2004) and Santos (2012), spaces like these are suitable for carrying out educational activities that can involve the community, such as conversation circles, workshops to exchange plant seedlings, and the valorisation of family herbal medicine. These initiatives are in line with public policies for sustainable rural development and solidarity (SILVA & MORAES, 2008).

6.10 ETHNOPHARMACOLOGICAL COLLECTION: MEDICINAL SPECIES USED

The ethnopharmacological survey inventoried a total of 131 species from 49 botanical families, 105 species and 41 botanical families in the MST group and 75 species and 32 botanical families in the settler group (Table 3).

Table 3: List of species mentioned by the groups studied (MST and Colonos), in alphabetical order of botanical families, followed by species, local vernacular nomenclature, origin, habit (Ar - arboreal; At - shrubby; Hb - herbaceous; Vi - vine), collection site (C - Cultivated, R - Ruderal, F - Forest) and voucher obtained from the Leopoldo Krieger Herbarium - CESJ.

Botanical Family	Species Name	Vernacular name		Place of Origin	Habit	Collection site	Voucher
		MST	Settlers				
Adoxacaeae	*Sambucus nigra* L.	Elderberries		Brazil	Air	R	
Alismataceae	*Echinodorus grandiflorus* (Cham. & Schltdl.) Micheli		Leather hat	Atlantic Forest	Hb	R	
Amaranthaceae	*Amaranthus viridis* L.	Cariru de Porco	Cariru de Porco	Caribbean	Hb	R	65048
	Dysphania ambrosioides (L.) Mosyakin & Clemants	St Mary's grass	St Mary's grass	Brazil	Hb	C	65064
	Beta vulgaris L.	Beetroot	Beetroot	Mediterranean	Hb	C	
Amaryllidaceae	*Allium cepa* L.	Head onion	Head onion	Asia	Hb	C	
	Allium ampeloprasum L.	Leek		Mexico	Hb	C	
	Allium fistulosum L.		Chives	Siberian Region	Hb	C	
	Allium sativum L.	Garlic		Asia	Hb	C	
Apiaceae	*Cyclospermum leptophyllum* (Pers.) Sprague	Macelinha		Atlantic Forest	Hb	R	65065
	Coriandrum sativum L.	Coriander		Brazil	Hb	C	
	Foeniculum vulgare Mill.	Fennel	Fennel	Mediterranean	Hb	C	

	Petroselinum crispum (Mill.) Fuss	Parsley	Parsley	Europe	Hb	C	
	Daucus carota L.	Carrot		Asia	Hb	C	
Araceae	*Xanthosoma sagittifolium* (L.) Schott.	Taioba		Atlantic Forest	Hb	C	
Aristolochiaceae	*Aristolochia* SP	Jug		Brazil	I saw	F	
Asparagaceae	*Agave americana* L.	Piteira		Mexico	Hb	C	
Bignoniaceae	*Pyrostegia venusta* (Ker Gawl.) Miers	Cipó São João		Atlantic Forest	I saw	F	
	Fridericia chica (Bonpl.) L.G. Lohmann	Indian grass		Atlantic Forest	At	C	
	Handroanthus impetiginosus (Mart. ex DC.) Mattos		Purple Ipe	Brazil	Air	F	
	Jacaranda caroba (Vell.) DC.	Courgette		Atlantic Forest	Air	F	
Bixaceae	*Bixa orellana* L.		Urucum	Atlantic Forest	Air	C	
Boraginaceae	*Symphytum officinale* L.	Comfrey		America	Hb	R	
	Tournefortia paniculata Cham.		Marmalade	Atlantic Forest	Hb	C	
Brassicaceae	*Nasturtium officinale* R.Br.	Watercress		Brazil	Hb	C	
	Brassica rapa L.	Sawdust		Atlantic Forest	Hb	C	
	Brassica oleracea L.	Cabbage	Kale	Mediterranean	Hb	C	
Bromeliaceae	*Ananas* sp.	Gravatá		Atlantic Forest	Hb	C	
Caricaceae	*Carica papaya* L.	Papaya		Caribbean	Air	C	
Asteraceae	*Artemisia absinthium* L.	Losna	Losna	Europe	Hb	C	
	Ageratum conyzoides (L.) L.		St John's	Atlantic Forest	Hb	R	
	Baccharis trimera (Less) DC		Carqueja	Brazil	Hb	R	
	Bidens pilosa L.	Prickly pear	Prickly pear	Tropical America	Hb	R	65037
	Solidago chilensis Meyen	Arnica		South America	Hb	R	65056
	Tithonia diversifolia (Hemsl.) A.Gray	Hand of God		India	Hb	R	65052
	Vernonanthura phosphorica (Vell.) H.Rob.	Fish tea	Fish tea	Brazil	Air	C	65047
	Calendula officinalis L.	Marigol		Europe	Hb	C	

		d					
	Matricaria chamomilla L.	Chamomile	Camomile	Europe	Hb	C	
	Tanacetum parthenium (L.) Sch.Bip.	Artemisia		Brazil	Hb	C	
	Mikania glomerata Spreng.	Guaco/ Guapo	Guaco/ Guapo	Atlantic Forest	I saw	F	
	Centaurea benedicta (L.) L.	Holy Thistle	Holy Thistle	Europe	Air	F	
	Chromolaena maximuiann (Schrad. Ex DC.) RMKing & H.Rob.	Arnica	Arnica	Atlantic Forest	Air	F	65041
	Acmella ciliata (Kunth) Cass.		Necroton	Atlantic Forest	Hb	C	
	Lactuca sativa L.	Lettuce	Lettuce	Meditarranean	Hb	C	
	Ageratum conyzoides (L.) L.	Mentraz		Tropical America	Hb	R	65039
	Achillea millefolium L.	Pronto Relief		Europe	Hb	C	
Convolvulaceae	*Ipomoea batatas* (L.) Lam.	Sweet potatoes	Sweet potatoes	South America	Hb	C	65062
Costaceae	*Costus spicatus* (Jacq.) Sw.	Monkey cane	Monkey cane	Brazil	Air	R	
Crassulaceae	*Kalanchoe laciniata* (L.) DC.	Skirt	Skirt	Brazil	Hb	C	
	Sedum dendroideum Moc. & Sessé ex DC.	Baspo	Baspo	Mexico	Hb	C	
Cucurbitaceae	*Momordica charantia* L.	São Caetano	São Caetano	Atlantic Forest	I saw	R	65049
	Cucumis anguria L.	Maxixe		India	Hb	C	
	Cucurbita moschata Duchesne	Pumpkin	Pumpkin	Asia	Hb	C	
	Sechium edule (Jacq.)SW.	Chuchu		Central America	I saw	C	
Euphorbiaceae	*Manihot esculenta* Crantz	Cassava		Atlantic Forest	Hb	C	
Lamiaceae	*Leonotis nepetifolia* (L.) R.Br.	Friar's string		Brazil	Hb	R	65043
	Leonurus sibiricus L.	Mané Turé/ Mané Magro/ Macaé	Mané Magro	Siberia	Hb	R	65053
	Peltodon radicans Pohl	Mint		Atlantic Forest	Hb	R	
	Plectranthus barbatus Andrews	Bilberry	Bilberry	India	At	C	65044
	Plectranthus	Awning		South	At	C	

	neochilus Schltr.			America			
	Glechoma hederacea L.	Grass Land		Macau	Hb	C	
	Lavandula latifolia Medik.	Lavender		Brazil	Hb	C	
	Melissa officinalis L.	Grass Cider		Pakistan	Hb	C	
	Mentha x piperita L.	Mint	Mint	India	Hb	C	
	Mentha arvensis L.	Vick		Europe	Hb	C	
	Mentha pulegium L.	Support	Support	Brazil	Hb	C	
	Mentha spicata L.	Black Mint		Brazil	Hb	C	
	Mentha viridis (L.) L.	Elevating		Brazil	Hb	C	
	Ocimum basilicum L.	Basil	Basil	India	Hb	C	
	Ocimum gratissimum L.	Lavender	Lavender	Asia	Hb	C	65046
	Ocimum carnosum (Spreng.) Link & Otto ex Benth.	Sweetgrass	Sweetgrass	Africa	Hb	C	
	Plectranthus amboinicus (Lour.) Spreng.	Mint Pepper		New Guinea	Hb	C	
	Origanum vulgare L.		Oregano	Europe	Hb	C	
	Mentha sp.		Elevating	Europe	Hb	R	
Lauraceae	*Persea americana* Mill.	Avocado	Avocado	Tropical America	Air	C	
	Laurus nobilis L.		Laurel	Brazil	Air	C	
	Cinnamomum verum J.Presl		Cinnamon	Sri Lanka	Air	C	
Fabaceae	*Anadenanthera colubrina* (Vell.) Brenan	Angico	Angico	Atlantic Forest	Air	C	
	Desmodium incanum DC.	Carrapichinha	Carrapichinha	Atlantic Forest	Air	F	65057
	Pterodon emarginatus Vogel	Sucupira		Atlantic Forest	Air	F	
Loranthaceae	*Struthanthus marginatus* (Desr.) G.Don		Birdsfoot trefoil	Atlantic Forest	Hb	C	
Lythraceae	*Punica granatum* L.		Pomegranate	Asia	Air	C	
Malpighiaceae	*Malpighia glabra* L.		Acerola	Americas	Air	C	
Malvaceae	*Gossypium hirsutum* L.	Cotton	Cotton	Brazil	Air	C	
	Hibiscus sp.	Grease fittings		Central America	At	C	
	Abelmoschus esculentus (L.) Moench	Okra	Okra	Africa	Hb	C	65058
Moraceae	*Morus* sp.	Mulberr		Asia	Air	C	

		y					
Musaceae	*Musa x. paradisíaca* L.		Banana	India	Hb	C	
Myrtaceae	*Psidium guajava* L.	Guava		America	Air	C	
	Syzygium cumini (L.) Skeels		Jamelão	India	Air	F	
Oleaceae	*Jasminum officinale* L.		Jasmine	Asia	Hb	C	
Passifloraceae	*Passiflora edulis* Sims	Passion fruit		Atlantic Forest	I saw	C	
Phyllanthaceae	*Phyllanthus niruri* L.	Breaking Stone	Breaking Stone	America	Hb	R	
	Phyllanthus tenellus Roxb.	Breaking Stone	Breaking Stone	America	Hb	R	65040
Phytolaccaceae	*Petiveria alliacea* L.		Guinea	Tropical America	Hb	C	65042
	Gallesia integrifolia (Spreng.) Harms	Pau d'alho		Atlantic Forest	Air	F	65050
Piperaceae	*Piper aduncum* L.	Jaborandí		Caribbean	At	F	65055
	Piper umbellatum L.	capeba		Atlantic Forest	Hb	F	
Plantaginaceae	*Plantago major* L.	Braiding	Braiding	Europe	Hb	R	65059
Poaceae	*Bambusa* sp.	Bamboo		Asia	Air	F	
	Arundo donax L.	Indian cane		Brazil	Air	C	
	Coix lacryma-jobi L.	Tear beads	Tear beads	Brazil	Hb	R	65054
	Cymbopogon citratus (DC.) Stapf	Grass Lemongrass	Lemongrass	India	Hb	R	65045
	Cymbopogon winterianus Jowitt ex Bor	Citronella grass	Citronella grass	Europe	Hb	C	
	Zea mays L.	Maize	Maize	Mexico	Hb	C	65060
Polygonaceae	*Persicaria hydropiper* (L.) Delarbre	Bug weed		North America	Hb	C	65038
Rosaceae	*Rosa alba* L.		White Rose	Mediterranean	At	C	
	Malus domestica Borkh.		Apple	Asia	Air	C	
Rubiaceae	*Genipa americana* L.	Jenipapo		Atlantic Forest	Air	F	
Rutaceae	*Citrus x aurantium* L.	Orange	Orange	India	Air	C	
	Citrus limon (L.) Osbeck	Lemon	Lemon	Asia	Air	C	
	Ruta graveolens L.	Rue	Rue	Southern Europe	Hb	C	
	Citrus aurantiifolia (Christm.) Swingle		Lima	Asia	Air	C	
Smilacaceae	*Smilax longifolia* Rich.	Sarsaparilla		South America	Hb	C	
Solanacea	*Nicotiana tabacum* L.	Smoke	Smoke	India	Hb	C	

e	*Solanum aethiopicum* L.	Jiló		Africa	Hb	C	
	Solanum tuberosum L.	English potatoes		Andes	Hb	C	
	Solanum cernuum Vell.	Panacea	Panacea	Atlantic Forest	Air	F	65063
	Lycopersicon esculentum Mill.		Cherry tomatoes	Tropical America	Hb	C	
	Solanum melongena L.		Aubergine	India	Hb	C	
	Solanum tuberosum L.		Potato	Andes	Hb	C	
	Solanum lycopersicum L.		Tomatoes	Americas	Hb	C	
Talinaceae	*Talinum paniculatum* (Jacq.) Gaertn.		Soft arse	Atlantic Forest	Hb	R	
Urticaceae	*Urtica dioica* L.	Nettle		Europe	Hb	R	
Verbenaceae	*Lantana camara* L.	Camará		Atlantic Forest	Hb	R	
	Stachytarpheta cayennensis (Rich.) Vahl	Gibbon		South America	Hb	R	
	Lippia alba (Mill.) N.E.Br. ex Britton & P.Wilson	Grass Lemon balm Twig		Atlantic Forest	At	C	65051
Lamiaceae	*Rosmarinus offtcinalis* L.	Rosemary	Rosemary	Pakistan	Hb	C	
Xanthorrhoeaceae	*Aloe vera* (L.) Burm.f.	Babosa	Babosa	Arabia	At	C	
Zingiberaceae	*Zingiber officinale* Roscoe	Ginger	Ginger	Asia	Hb	C	
	Hedychium coronarium J.KOENIG		Imbirí	Atlantic Forest	Air	F	

Based on the results presented, 37.13 per cent similarity was obtained between the species used by the two groups (MST and Colonos), highlighting the greater knowledge of the MST group.

It was observed that of the species used in common by both groups, the vernacular names coincided completely, despite the fact that the members of the MST and Colonos come from different backgrounds, as already mentioned in the discussion of item 6.1 (socio-cultural data), and that they may have suffered cultural influences from common origins. Popular names are fundamental in the communication between popular knowledge and the botanical nomenclature used by scientific methodology, so access to and referencing of medicinal species must move towards standardising popular terminology and be validated by scientific methodology, as used in the work of Tomazi (2014), who states that vernacular names are useful and important in ethnobotanical research, as the origin of references about the culture or vocabulary of a population, and can provide clues as to the popular use of a particular species. In this way, ethnobotanical studies make it possible to integrate the vernacular names produced by a given population with scientific knowledge about wisdom and natural events (STRACHULSKI, 2013). Currently, the use of medicinal plants in Brazil is influenced by the various cultures received from indigenous peoples, colonising Europeans, African slaves and immigrants (European and Asian) who arrived in Brazil after the Second World War (PEREIRA, 2002). Almeida (2003) emphasises that in the South and Southeast of Brazil, medicinal plants of European origin are the most expressive, which coincides with the migrations from Europe that prevailed in these regions. This can be explained by the historical influence in Brazil of the coexistence of European, African and Asian cultures in the treatment of diseases (ALMEIDA, 2000). However, exotic species can have a negative impact on the local biome, as they destabilise ecosystems, habitats or species native to a given environment (BRASIL, 2002). Although these exotic species may occupy the environment naturally, they cause a transformation in the structure and composition of natural systems, eliminating the peculiarly natural characteristics of biodiversity in a given location (SIQUEIRA, 2014). This should imply prioritising the choice of plants to be used in the medicinal garden to be set up and discussed in item 6.9. In the case of the groups (MST and Colonos), these plants are predominantly cultivated in their gardens, so they are a renewable resource and corroborate the practices of encouraging sustainability. Furthermore, they do not compromise the biodiversity of the local native flora, since once cultivated, they are offered as an alternative resource and raw material instead of using native species, which becomes a fact related to sustainability.

Of all the species inventoried, 22.73 per cent are native to the Atlantic Forest biome. Of the exotic species (77.27%), the main ones are Asian (18.19%) and European (9.85%). According to Conde (2016), the lower number of native plants found in this study may be related to the reduced degree of isolation of the populations in question, since although the members of the Colonos have been in the area for several generations, they have been influenced by various cultures, such as European, Asian and African. However, with regard to the MST group, who also come from rural Minas Gerais, the predominant use of exotic species may be related to the diverse cultural influences that rural populations in the state have historically suffered (FARIAS, 2016).

In terms of habit, herbaceous plants are the most used, which could indicate that the form of management in the groups (MST and Colonos) would be defined as low impact, since exotic and herbaceous species are being used prominently (CONDE **et al**. 2014). This result could suggest that the management practised in the communities is being characterised as low impact, since exotic and herbaceous species are predominantly used (CONDE **et al.** 2014). The preference for the herbaceous habit may be related to the availability and easy accessibility of herbaceous plants used in homemade preparations, most of which are found in backyards and vegetable gardens (SIQUEIRA, 2014). However, tree species are also widely used, but herbaceous species are more widely used because they are easier and quicker to grow, as observed by Hanazaki (2006); Pinto **et al.** (2006) and Di Stasi **et al**. (2002).

Regarding the collection sites (occurrence) of the species surveyed for both groups, 67.17 per cent are cultivated, 19.80 per cent are ruderal and 12.90 per cent are forested. According to Vieira & Silva (2002), the cultivation of a particular species can end up interfering positively in its perpetuation, thus serving as a conservation strategy for it. However, most of the species used have exotic origins and have been cultivated on site, which according to Conde (2016), can spare forest and native species from possible degradation.

In the MST group, the predominant botanical families were: Lamiaceae (18 species), Asteraceae (14 species), Poaceae (6 species), Curcubitaceae (4 species), Solanaceae (4 species), Amaranthaceae (3 species), Amaryllidaceae (3 species), Bignoniaceae (3 species), Brassicaceae (3 species), Fabaceae (03 species), Malvaceae (3 species), Phytolaccaceae (3 species), Rutaceae (03 species) and Verbenaceae (3 species), which corresponds to other studies carried out in Brazil, such as Brito and Brito (1993) and Maioli- Azevedo and Fonseca-Kruel (2007). The botanical families that predominated in the Settlers group were: Asteraceae (11 species), Lamiaceae (10 species), Solanaceae (6 species), Poaceae (04 species), Rutaceae (04 species), Amaranthaceae (3 species), Amaryllidaceae (03 species) and Lauraceae (03 species).

It was possible to see that between the two groups (MST and Colonos) the Lamiaceae and Asteraceae families were predominant, which according to Carvalho **et al**. (2013), may be related to the fact that these families are cosmopolitan, adapting to the most diverse types of environments. The predominance of these botanical families was also observed in works by Vendruscolo & Mentz (2006); and Pinto & Amorozo (2006). According to Bennett & Prance (2000), the Lamiaceae and Asteraceae families are commonly used for the production of medicines, and are found in both tropical and temperate climates, leading the group of introduced medicinal plants.

6.11 NATIVE SPECIES IN THE ETHNOPHARMACOLOGICAL COLLECTION AND THEIR CURRENT CONSERVATION STATUS

With regard to the native Atlantic Forest species used by the MST, none of them appear in the manuals: IUCN - Red List of Threatened Species (2014); Ministry of the Environment (2008); Biodiversitas Foundation (2009). The Colonos group also found no native Atlantic Forest species in the manuals: IUCN - Red List of Threatened Species (2016); Ministry of the Environment (2008); Biodiversitas Foundation (2009).

The fact that none of the Atlantic Forest's native species appear on lists of threatened species reduces the risks to the environment and may be related to the increase in cultivation and use of exotic species (BENNETT & PRANCE 2000).

Based on these results and CNC-Flora (2016), it is possible to see that the use of species from ethnopharmacological knowledge has little impact on local extractivism. It is not possible to detect vulnerability of the species inventoried in terms of their uses, for any of the groups studied (MST and Settlers). However, the importance of implementing initiatives aimed at local environmental education should be emphasised, so that future problems related to extractivism in the area do not occur (CONDE **et al.** 2014).

6.12 VULNERABILITY OF NATIVE MEDICINAL SPECIES IN THE CONTEXT OF LOCAL ETHNOPHARMACOLOGICAL KNOWLEDGE

As for the medicinal species that are at risk of use, only Category 3 species occur in the MST group, while two Category 2 species occur in the Settlers' group and the rest are Category 3 (Table 4 and 5).

Based on the results, it can be seen that among the MST group, all the species are in Category 3. In the Settlers group, the species *Brassica rapa; Solidago chilensis;* and *Chromolaena maximilianii are in* category 2 and the others are in category 3 in terms of priority for conservation. As such, although these species can only be moderately collected, they do not present any risks in terms of their utilisation, since they are annual and ruderal herbaceous plants with high environmental resilience and wide dispersal.

Table 4: List of native species in the MST group, followed by their vernacular names, number of citations, number of uses, use value (VU), part used (CA - bark; FL - flower; FO - leaf; RZ - root; S - seed; PT - whole plant), collection risk (C), diversity of use attributed to the species (Div), local use (L) and utilisation risk (RU).

Species Name	Popular Name	No. of Quote	N° Uses	VU	Part Used	Collection Risk [C]	Div	Local Use [L]	UK	[(VU x 10) + RU] / 2
Peltodon radicans Pohl	Mint	1	1	0,1	EN	10	1	1	55	28
Pterodon emarginatus Vogel	Sucupira	2	1	0,1	CA; FO; RZ	10	1	1	55	28
Lantana camara L.	Camará	2	3	0,3	FO	1	3	1	40	21,5
Solanum cernuum Vell.	Panacea	2	3	0,3	FO	1	3	1	40	21,5
Anadenanthera colubrina (Vell.) Brenan	Angico	5	2	0,2	FO	1	2	4	25	13,5
Chromolaena maximilianu (Schrad. ex DC.) R.M.King & H.Rob.	Arnica	5	2	0,2	FO	1	2	4	25	13,5
Lippia alba (Mill.) N.E.Br. ex Britton & P.Wilson	Lemon Balm	3	2	0,2	FO	1	2	4	25	13,5
Mikania glomerata Spreng.	Guaco	5	2	0,2	FO	1	2	4	25	13,5
Pyrostegia venusta (Ker Gawl.) Miers	Cipó São João	1	2	0,2	FL	1	2	1	25	13,5
Gallesia integrifolia (Spreng.) Harms	Pau d'alho	2	1	0,1	CA	4	1	1	25	13
Manihot esculenta Crantz	Cassava	3	1	0,1	FO	1	1	4	25	13
Momordica charantia L.	São Caetano	3	1	0,1	FO	1	1	4	25	13

Cyclospermum leptophyllum (Pers.) Sprague	Macelinha	1	1	0,1	FO	1	1	1	10	5,5
Desmodium incanum DC.	Carrapichinha	1	1	0,1	FO	1	1	1	10	5,5
Fridericia chica (Bonpl.) L.G. Lohmann	Indian grass	1	1	0,1	FO	1	1	1	10	5,5
Genipa americana L.	Jenipapo	1	1	0,1	FO	1	1	1	10	5,5
Jacaranda caroba (Vell.) DC.	Courgette	2	1	0,1	FO	1	1	1	10	5,5
Piper umbellatum L.	capeba	1	1	0,1	FO	1	1	1	10	5,5

Table 5: List of the native species in the Colonos group, followed by their vernacular names, number of citations, number of uses, use value (VU), part used (CA - bark; FL - flower; FO - leaf; RZ - root; S - seed; PT - whole plant), collection risk (C), diversity of use attributed to the species (Div), local use (L) and utilisation risk (RU).

Species Name	Name Popular	N° Quote	No. of Uses	VU	Part Used	Risk of Collection [C]	Div	Local Use [L]	UK	[(VU x 10) + RU] / 2
Brassica rapa L.	Sawdust	1	1	0,36	EN	10	5	4	75	39,3
Solidago chilensis Meyen	Arnica	4	2	0,21	EN	10	3	4	70	36,05
Chromolaena maximilianii (Schrad. ex DC.) R.M.King & H.Rob.	Arnica	5	3	0,07	EN	10	1	4	70	35,35
Desmodium incanum DC.	Carrapichinha	3	1	0,07	EN	10	1	1	55	27,85
Hedychium coronarium J.KOENIG	Imbirí	1	1	0,07	RZ	10	1	1	55	27,85
Ageratum conyzoides (L.) L.	St John's	1	1	0,07	EN	10	1	1	55	27,85
Acmella ciliata (Kunth) Cass.	Necroton	4	5	0,28	FO	1	4	4	25	13,9
Phyllanthus tenellus Roxb.	Breaking Stone	6	1	0,21	FO	1	3	4	25	13,55
Phyllanthus niruri L.	Breaking Stone	5	1	0,14	FO	1	2	4	25	13,2
Momordica charantia L.	São Caetano	5	4	0,07	FO	1	1	4	25	12,85
Passiflora edulis Sims	Passion fruit	1	1	0,14	FR	1	2	1	15	8,2
Anadenanthera colubrina (Vell.) Brenan	Angico	1	1	0,07	FO	1	1	1	10	5,35
Talinum paniculatum (Jacq.) Gaertn.	Soft arse	1	1	0,07	FO	1	1	1	10	5,35
Echinodorus grandiflorus (Cham. & Schltdl.) Micheli	Leather hat	1	1	0,07	FO	1	1	1	10	5,35
Struthanthus marginatus (Desr.) G.Don	Birdsfoot trefoil	3	3	0,07	FO	1	1	1	10	5,35

Ananas sp.	Gravatá	1	1	0,07	FR	1	1	1	10	5,35
Mikania glomerata Spreng.	Guaco	1	1	0,07	FO	1	1	1	10	5,35
Tournefortia paniculata Cham.	Marmalade	1	1	0,07	FO	1	1	1	10	5,35
Solanum cernuum Vell.	Panacea	2	2	0,07	FO	1	1	1	10	5,35
Xanthosoma sagittifolium (L.) Schott.	Taioba	1	1	0,07	FO	1	1	1	10	5,35
Bixa orellana L.	Urucum	1	1	0,07	S	1	1	1	10	5,35

6.13 THE USE OF MEDICINAL PLANTS AND ANVISA'S RECOMMENDATIONS (RDC No. 10), AND THE SCIENTIFIC LITERATURE

With a view to setting up a community medicinal garden with the appreciation of local ethnopharmacological knowledge and the safe use of the species listed, it can be seen that among the MST group, 39 species had their uses supported by scientific literature and 12 by ANVISA. Among the Settlers group, 41 species had their uses supported by scientific literature and 15 by ANVISA. However, for the MST, of these species whose uses were supported by ANVISA, 10 are used in complete agreement, but 1 (**Phyllanthus niruri**) did not coincide in terms of the part used, and 1 (**Calendula officinalis**) did not coincide in terms of the method of preparation. In the Colonists' group, 15 species are used completely in accordance with ANVISA, but of the species whose uses were supported by ANVISA, 1 **Phyllanthus niruri did** not coincide in terms of the part used (Table 6).

Table 6: Species cited by the MST and Colonos groups, followed by their vernacular names, part used (Bu - bulb; Ca - bark; Cl - stem; Fl - flower; Fo - leaf; Fr - fruit; Rz - root; Se - seed; Pt - whole plant), method of preparation (Co - edible; Dc - decoction; In - infusion; Ma - maceration; Po - ointment; Ti - tincture; Xa - syrup), main use, pharmacological validation and citation of the plant by RDC-10/ANVISA; where (Y) means, yes or used for those used totally in line with ANVISA, and (N), means, no, or for those used for the same recommended therapeutic purpose, but which have a different part used or method of preparation.

Species Name	Part Used		Preparation method		Main use		Pharmacological validation		ANVISA (Y) or (N)	
	MST	Settlers	MST	Settlers	MST	Settlers	MST	Settlers	MST	Settlers
Abelmoschus esculentus	Fr	Fr	Co	Co	Diabetes	High blood pressure	Antidiabetic (Sabitha *et al.* 2011)			
Achillea millefolium	Fo		In		Pain		Antinociceptive (Pires *et al.* 2009)		X/ (S)	
Acmella ciliata		Fo		Ma		Liver and stomach	Gastroprotective (Boeing *et al.* 2016)			

							"			
Agave americana	Fl		Ma		Pain			Anti-inflammatory (Monterrosas *et al.* 2013)		
Ageratum conyzoides	Fo	Pt	In	Dc	Anti-inflammatory	Abortive	Antinociceptive (Hossain *et al.* 2013)		X/ (S)	
Allium ampeloprasum	Fo		Ma		Flu					
Allium cepa	Bu	Ca	In	In	Flu	Stomach	Antigripal (Passalacqua *et al.*. 2007)			
Allium fistulosum		Fo		Co		Cracked feet				
Allium sativum	Bu		Ma		Ear pain		Anti-inflammatory (Ndoye *et al.* 2016)			
Aloe vera	Fo	Fo	Ma	Ma	Cancer	Hair loss; Cancer	Anti-neoplastic (Hussain *et al.* 2015)	Anti-neoplastic (Hussain *et al.* 2015)		
Amaranthus viridis	Fo	Fo	In	In	Increase lactation	Increase lactation; Anaemia				
Anadenanthera colubrina	Fo	Fo	Xa	In	Flu	Throat infection	Antinociceptive (Damascena et al. 2014), Anti-inflammatory (Santos *et al.* 2013)	Antinociceptive (Damascena et al. 2014), Anti-inflammatory (Santos *et al.* 2013)		
Ananas sp.	Fr		Xa		Flu					
Aristolochia sp.	Rz		Dc		Fever					
Artemisia absinthium	Fo	Fo	In	Ma	Stomach pain and liver problems	Stomach pain	Antiulcerogenic (Shafi *et al.* 2004); Anti-inflammatory and Antinociceptive (Zeraati *et al.* 2014); Hepatoprotective (Amat *et al.* 2010)	Antiulcerogenic (Shafi *et al.* 2004); Anti-inflammatory and anti-nociceptive (Zeraati *et al.* 2014)		
Arundo donax	Fo		In		Infection					
Baccharis trimera		Pt		In		High cholesterol		Hypocholesterolaemic (Souza, 2012)		
Bambusa sp.	Cl		Ma		Digestion					
Beta vulgaris	Rz	Rz	Co	Xa	Anaemia	Coughing	Anti-anaemic (Indhumathi & Kannikaparameswari, 2012)	Expectorant (Taur *et al.* 2010)		

Bidens pilosa	Fo	Fo	In	In	Icteric	Liver	Hepatoprotective (Yuan *et al.* 2008)	Hepato-protective (Yuan *et al.*. 2008)	X/ (S)	X/ (S)
Bixa orellana		If		Ma		High blood pressure		Diuretic/hypotensive (Radhika *et al.* 2010)		
Brassica oleracea	Fo	Fo	Ma	Ma	Anaemia; hair loss	Stomach		Antiulcerogenic(Jiva & Bahmani,2016.)		
Brassica rapa	Pt		Co		Blood purifying					
Calendula officinalis	Fl		In		Healing		Healing (Okuma *et al.* 2015)		X/ (N)	
Carica papaya	Fo		Dc		Congestion		Stomachic (Muss, 2013)			
Centaurea benedicta	Fo	Fo	In	In	Pneumonia	Headache				
Dysphania ambrosioides	Fo		In		Vermifuge		Anthelmintic (MacDonald *et al.* 2004)			
Chromolaena maximilianii	Fo	Pt	Ti	Ma; Ti	Inflammation	Pain caused by knocks				
Ciclospermum leptophyllum	Fo		Ma		Belly ache					
Cinnamomum verum		Cl		Dc	Liver ailments			Antioxidant (Dhuley, 1999)	X/ (S)	
Citrus aurantiifolia		Fr		Co		Kidneys				
Citrus aurantium	Fo	Fo	In	In	Flu	Flu	Antinociceptive and anti-inflammatory (Khodabakhsh *et al.* 2015)	Antinociceptive and anti-inflammatory (Khodabakhsh *et al.* 2015)		
Citrus limon	Fo	Fr	Ma	Dc	Flu-like symptoms	Flu-like symptoms	Antinociceptive (Campêlo *et al.* 2011)	Antinociceptive (Campêlo *et al.* 2011)		
Coix lacryma-jobi	Fo; Rz; Se	If	In	Dc	Flu	Kidneys				
Coriandrum sativum	Fo; Rz; Se		In		Colic					
Costus spicatus	Cl; Fo	Fo	In; Dc	In	Kidney pain	Kidney pain	Antinociceptive and anti-inflammatory (Quintans Júnior *et al.* 2010)	Antinociceptive and anti-inflammatory (Quintans Júnior *et al.* 2010)		
Cucumis anguria	Fr		Co		Anaemia					
Cucurbita moschata	Cl	Fo	Dc	In	Colic	Ear pain		Anti-inflammatory		

								(Lee *et al.* 2015)		
Cymbopogon citratus	Fo	Fo	In	In	Flu	Soothing	Antigripal (Oboh *et al.*. 2010)	Anxiolytic (Costa *et al.* 2011)		X/ (S)
Cymbopogon winterianus	Fo	Fo	In	Ti	Repellent	Repellent	Acaricide (Mello *et al.* 2014)	Acaricide (Mello *et al.* 2014)		
Daucus carota	Rz		Co		Dry skin					
Desmodium incanum	Fo	Pt	In	In	Kidney problems	Kidney problems				
Dysphania ambrosioides		Fo		Dc		Vermifuge		Anthelmintic (MacDonald *et al.* 2004)		
Echinodorus grandiflorus		Fo		In		Kidney problems				
Foeniculum vulgare	Pt	Pt	In	In	Belly ache	Belly ache	Analgesic and anti-inflammatory (Elizabeth *et al.*. 2014)	Analgesic and anti-inflammatory (Elizabeth *et al.* 2014)		
Fridericia chica	Fo		In		Belly ache		Antinociceptive (Michel *et al.* 2015)			
Gallesia integrifolia	Ca		Dc		Column					
Genipa americana	Fo		In		Joint pain					
Glechoma hederacea	Fo		Xa		Flu					
Gossypium hirsutum	Fo; Se	Fo; Se	In; Ma	In; Ma	Throat infection	Uterine infection; Earache	Antimicrobial (Miranda *et al.* 2013)			
Handroanthus impetiginosus		Fo		Dc		Healing		Antimicrobial (Vasconcelos *et al.* 2014)		
Hedychium coronarium		Rz		Dc		Toothache		Analgesic and anti-inflammatory (Shrotriya *et al.* 2007)		
Hibiscus sp.	Fl		In		Hair loss		Anti-hair loss (Patel *et al.* 2015)			
Ipomea batatas	Fo	Fo	Ma	Dc	Anaesthetic for toothache	Anaesthetic for toothache				
Jacaranda caroba	Fo		Ma		Anti-inflammatory					
Jasminum officinale		Fo		In		Diabetes				
Kalanchoe laciniata	Fo	Fo	In	Co	Stomach problems	Ulcers	Anti-inflammatory (Mourão *et al.* 1999)	Anti-inflammatory (Mourão *et al.* 1999)		

Lactuca sativa	Fo	Pt	Co	In	Insomnia	Soothing	Sedative (González-Lima *et al.* 1996)	Sedative (González-Lima *et al.* 1996)		
Lantana camara	Fo		Ma		Flu		Analgesic and anti-inflammatory (Silva *et al.* 2015)			
Laurus nobilis		Fo		In		Flu		Analgesic and anti-inflammatory (Sayyah *et al.* 2003); " Antimicrobial (Hussaini *et al.* 2009)		
Lavandula angustifolia	Cl		In		Belly ache					
Leonotis nepetifolia	Fo		In		Stomach			Analgesic and anti-inflammatory (Pushpan *et al.* 2012)		
Leonurus sibiricus	Fo	Fo	In	In	Stomach	Belly ache	Analgesic and anti-inflammatory (Islam *et al.* 2005);	Analgesic and anti-inflammatory (Islam *et al.*. 2005); Antidiarrhoeal (Almeida *et al.*. 2005)		
Lippia Alba	Fo		In		Soothing		Sedative (Zétola *et al.*. 2002)		X/ (S)	
Lycopersicon esculentum		Fo		Dc		Anti-inflammatory		Anti-inflammatory (Li *et al.* 2014)		
Malpighia glabra		Fr		Co		Flu				
Malus domestica		Fr		Ma		Insomnia				
Manihot esculenta	Fo		Ma		Anaemia					
Matricaria recutita	Fo	Fl	In	In	Fever	Belly ache				X/ (S)
Melissa officinalis	Pt		Xa		Flu					
Mentha arvensis	Fo		In		Flu					
Mentha piperita	Fo	Fo	In	In	Belly ache	Vermifuge; Soothing; Colic Menstrual	Antinociceptive (Taher, 2012)	Antiparasitic (Naranjo *et al.* 2006); Antinociceptive (Taher, 2012)	X/ (S)	X/ (S)
Mentha pulegium	Fo	Fo	In	In	Flu	Flu	Anti-inflammatory (Erhan *et al.* 2012)	Anti-inflammatory (Erhan *et al.*2012	X/ (S)	X/ (S)

Mentha sp.		Fo		In		Flu				
Mentha spicata	Fo		In		Flu					
Mentha viridis	Fo		Xa		Flu					
Mikania glomerata	Fo	Fo	Xa	In	Flu	Coughing	Bronchidilator (Leal *et al.* 2000) and expectorant (Leal *et al.* 2000; Soares de Moura *et al.* 2002)	Bronchidilator (Leal *et al.* 2000) and expectorant (Leal *et al.* 2000; Soares de Moura et al. 2002)	X/ (S)	X/ (S)
Momordica charantia	Fo	Fo	In	In	Dengue fever	Fever; Flu	Immunomodulator (Deng *et al.* 2014); possible inhibition of dengue virus adsorption and replication (Tang *et al.* 2012); Larvicide (Nagappan *et al.* 2014).	Immunomodulator (Deng *et al.* 2014); possible inhibition of dengue virus adsorption and replication (Tang *et al.* 2012); Influenza antiviral (Pongthanapisith *et al.* 2013)		
Morus sp.	Fo		In		Headache					
Paradise muse		Fr		Ma		Haemorrhoids				
Nasturtium officinale	Fo		In		Flu		Anti-inflammatory (Sadeghi *et al.*. 2014)			
Nicotiana tabacum	Fo	Fo	Ma	Ma	Healing	Baby's navel healer				
Ocimum basilicum	Fo	Fo	Ma	In	Digestive	Soothing	Hepatoprotective (Abd El-Salam *et al.*. 2015)	Anxiolytic (Gradinariu *et al.* 2014)		
Ocimum carnosum	Fl	Pt	In	In	Belly ache	Belly ache	Antispasmodic (Souza *et al.* 2015)	Antispasmodic (Souza *et al.* 2015)		
Ocimum gratissimum	Fo	Fo	In	In	Flu	*Flu*	Antinociceptive, anti-inflammatory (Tanko *et al.* 2008); Antinociceptive (Rabelo *et.al.* 2003); Antinociceptivo (Paula-Freire *et al.* 2012).			
Origanum vulgare		Pt		In		Coughing				
Passiflora edulis	Fo		In		Soothing		Andidepressant (Wang *et al.*		X/ (S)	

							2013); Anxiolytic (De-Paris *et al.* 2002)			
Peltodon radicans	Pt		Xa		Flu					
Persea americana	Fo	Fo	In	In	Kidney problems	Kidney problems	Urinary Tract Infection (Hernández *et al.*. 2014)	Urinary Tract Infection (Hernández *et al.* 2014)		
Petiveria alliacea		Fo		In		Unloading				
Petroselinum crispum	Fo	Fo	Po	In	Insomnia; Gastritis	Healing				
Phyllanthus niruri	Fo	Pt	In	In	Kidney problems	Kidney problems	Analgesic and anti-inflammatory (Santos *et al.* 1994); Antimicrobial (Ignácio *et al.* 2001), Antibiotic and anti-inflammatory for urinary tract (Khare *et al.* 2014)	Analgesic and anti-inflammatory (Santos *et al.* 1994); Antimicrobial (Ignacio *et al.* 2001), Antibiotic and anti-inflammatory of the urinary tract (Khare *et al.* 2014)	X/ (N)	X/ (N)
Phyllanthus tenellus	Pt	Pt	In	In	Kidney problems	Kidney problems	Antimicrobial (Ignácio et al. 2001), Analgesic and anti-inflammatory (Santos *et al.* 1994)	Antimicrobial (Ignácio *et al.* 2001), Analgesic and anti-inflammatory (Santos *et al.* 1994)		
Piper aduncum	Rz		Ma		Toothache; hair loss		Antimicrobial (Abreu *et al.* 2015); Anti-inflammatory (Parise- Filho *et al.* 2011)			
Piper umbellatum	Fo		In		Liver problems		Hepato-protective, cardio-protective and renal-protective (Agbor *et al.* 2012)			
Plantago major	Fo	Fo	In	In	Throat infection	Throat infection	Antibacterial (Karima *et al.* 2015)	Antibacterial (Karima *et al.* 2015)	X/ (S)	X/ (S)
Plectranthus barbatus	Fo	Fo	In	Ma	Stomach pain	Liver problems		Hepato - protective (Perandin *et al.* 2015)		X/ (S)
Plectranthus amboinicus	Fo		Xa		Bronchitis			Antibacterial (Paulo *et al.* 2009)		

Plectranthus neochilus	Fo		Ma		Stomach ache					
Persicaria hydropiper	Fo		In		Haemorrhoids					
Psidium guajava	Fo		In		Belly ache			Antidiarrhoeal (Mazumdar *et al*. 2015 and Shah *et al*. 2011)		X/ (S)
Pterodon emarginatus	Ca; Fo; Rz		In		Joint pain		Analgesic (Negri *et al*. 2014), Immune-inflammatory control (Alberti *et al*. 2014)			
Punica granatum		Fr		Ma		Throat infection		Antinociceptive and anti-inflammatory (González-Trujano *et al*. 2015)	X/ (S)	
Pyrostegia venusta	Fl		In		Bronchitis; flu		General anti-inflammatory (Veloso *et al*. 2014)			
Rosa Alba		Fl		In		Uterine infection				
Rosmarinus officinalis	Fo	Fo	In	In	Soothing	Dermatitis		Antidepressant (Sasaki *et al*. 2013;Machado, *et al*. 2009)	X/ (S)	X/ (S)
Ruta graveolens	Fo	Fo	Ti	Ma	The evil eye	Lice infestation		Pediculicide (Jorge *etal*. 2009)		
Sambucus nigra	Fl		In		Flu				X/ (S)	
Sechium edule	Cl		In		Hypertension			Antihypertensive (Lombardo-Earl *et al*. 2014)		
Sedum dendroideum	Fo	Fo	Ma	Ma	Ulcers	Cicatrizante		Antinociceptive and anti-inflammatory (Melo *et al*. 2009)		
Smilax longifolia	Rz		Dc		Anti-inflammatory					
Solanum aethiopicum	Fr		Ti		Pain					
Solanum cernuum	Fo	Fo	In	Dc	Blood purifying	Menstrual cramps; Itching		Analgesic and anti-inflammatory (Lopes *et al*. 2014)		

Solanum lycopersicum		Fo		In		Toothache		Antinociceptive and anti-inflammatory (Nascimento *et al.* 2016)		
Solanum melongena		Fr		Ma		High cholesterol		Cholesterol reducer (Cherem *et al.* 2007)		
Solanum tuberosum	Cl; Rz	Fo	Co	In	Stomach pain	Toothache	Antinociceptive and anti-ulcerogenic (Lee *et al.* 2009)			
Solidago chilensis	Pt	Pt	Ti	Ti	Inflammation	Pain caused by knocks	Anti-inflammatory (Tamura *et al.*. 2009; Goulart *et al.* 2007; Liz *et al.*. 2008), Analgesic (Silva *et al.*. 2015)	Anti-inflammatory (Tamura *et al.* 2009; Goulart *et al.* 2007; Liz *et al.* 2008), Analgesic (Silva *et al.* 2015)		
Stachytarpheta cayennensis	Fo		Dc		Gum inflammation					
Struthanthus marginatus		Fo		Ma		Pneumonia				
Symphytum officinale	Fo		In		Stomach pain					
Syzygium cumini		Fo		In		Diabetes		Anti-diabetic (Alam *et al.* 2012 and Srivastava *et al.* 2013)		
Talinum paniculatum		Fo		Co		Blood purifying				
Tanacetum parthenium	Fo		Ma		Joint pain		Anti-inflammatory (Sur *et al.* 2009)			
Tithonia diversifolia	Fl		Ti		Removing addictions					
Tournefortia paniculata		Fo		In		Kidneys				
Dioecious urtica	Fo		Dc		Haemorrhoids; erysipelas					
Vernonanthura phosphorica	Fo	Fo	In	In; Ma	Pneumonia	Flu	Anti-inflammatory (silva *et al.* 2012)	Anti-inflammatory (silva *et al.* 2012		
Xanthosoma sagittifolium	Fo		Co		Blood purifying					

Zea mays	Fl; Fr	Fl	In; Co	In	Bladder problems; malnutrition	Kidney stones		Anti-lithic (Rathod *et al.* 2013)		
Zingiber officinale	Rz	Rz	In	Co	Flu; soothing	Flu	Anti-inflammatory (Han *et al.* 2013 and Malipatil *et al.* 2015)	Anti-inflammatory (Han et al. 2013 and Malipatil *et al.* 2015)	X/ (S)	X/ (S)

Of the 95 medicinal species used by the MST group, 50 were backed up by scientific literature. Among the settlers, 53 had their uses backed up. However, the species whose uses have not been supported either in scientific literature or in RDC No. 10 - ANVISA, can be seen as potential for further pharmacological studies based on local ethnopharmacological knowledge (CONDE, 2012).

With regard to local health demands, it was found that the medicinal plants used cater for both groups (MST and settlers), since they can be used efficiently to treat the main health problems affecting local residents.

In the case of the MST group, the most cited illness was flu, followed by hypertension. The respective species used to treat these illnesses were Anadenanthera colubrina (Angico) and Mikania glomerata (Guaco). In the case of the Colonos group, the most cited illness was diabetes, followed by flu and hypertension. The species used to treat these illnesses were Jasminum officinale (Jasmine), Syzygium cumini (Jamelão) and Abelmoschus esculentus (Okra), Bixa orellana (Urucum), Sechium edule (Chayote) and Zingiber officinale (Ginger). Of these species, only *Abelmoschus esculentus'* main use for "high blood pressure" is not supported in scientific literature, which demonstrates the great relevance of ethnopharmacological knowledge in relation to local health demands. In this way, this knowledge can be recognised as being of great value as an alternative for treating health problems with sustainability and respect for the environment. According to Dias (2002), knowledge about the use and cultivation of medicinal plants in communities is a significant medicinal resource and can add sustainability to local development.

Based on these results, this study predicts the safe use of 20 species in relation to ANVISA RDC No. 10 (Table 7). However, some species that were not found on the ANVISA list have had their uses supported by scientific studies *(in vitro* and *in vivo* - pre-clinical), and could possibly be included on the list in the future if they are included in new ANVISA resolutions.

Table 7: List of species listed for priority planting in the local community garden, and whose local uses have been backed by ANVISA - RDC No. 10:

Achillea millefolium Ageratum conyzoides Bidens pilosa Calendula officinalis Cinnamomum verum Cymbopogon citratus Lippia alba	*Matricarici recutita Mentha piperita Mentha pulegium Mikania glomerata Passiflora edulis Phyllanthus niruri Plantago major*	*Plectranthus barbatus Psidium guajava Punic granatum Rosmarinus officinalis Sambucus nigra Zingiber officinale*

6.14 COMPARATIVE PARAMETERS BETWEEN THE GROUPS STUDIED

With regard to the summary of the comparative and contextualised assessment of the groups studied, it can be seen that there are a number of differences and similarities (Table 8), and these characteristics can serve as a basis for developing initiatives aimed at cultural aggregation and maintenance.

Table 8: Main comparative parameters in relation to the data obtained from the groups studied.

Parameters	**Comment**
Sample group and socio-cultural data	Interviewees: 13 MST and 14 settlers. Both had similar characteristics, but there was disagreement about the need for greater aggregation among the Colonos, with a view to reducing youth dropout, and integration between the groups.
Gender of those who get sick the most and most cited illnesses	Males are more affected by diseases in both groups, as are children and the elderly. The most frequently cited diseases were: MST, flu and hypertension; settlers: diabetes, flu and hypertension.
Learning and passing on knowledge	Both groups showed multiple
about Medicinal Plants	The MST group's knowledge is passed on effectively (100 per cent). However, among the settlers, this may be being jeopardised, as only 50% of the older settlers pass on their knowledge to the younger ones, who are leaving the area.
The use of industrialised medicines and medicinal plants.	0 he MST group predominantly uses medicinal plants, while the Colonos group predominantly uses industrialised medicines. Health sustainability and self-management are more promising in the MST group.
Access to the Unified Public Health System (SUS).	All the two groups compared have access to the SUS and use it.
Future prospects: building a medicinal garden.	All the participants in both groups approve of building a community medicinal garden.
Medicinal species used in terms of their habits, origins and predominance of botanical families.	Both groups use predominantly herbaceous species of exotic origin, most of which are cultivated. The main botanical families were Lamiaceae and Asteraceae.
Plant species native to the Atlantic Forest that are	None were found among the threatened species.

listed in the manuals of threatened species.	
Quantitative analyses and vulnerability of plant species.	In both groups, no species were found that were vulnerable in terms of their use and management.
Ethnopharmacological knowledge and the recommendations of RDC No. 10 - ANVISA	Of the species used by both groups (132), 20 were listed with support for their use (Table 7), the way they are prepared and the botanical part they use. They are therefore suitable for use in a community medicinal garden.

It was clear that there are traces of conflict in the interaction between the two groups (MST and settlers). These conflicts are justified by the fact that the MST group is settling in the area that was previously occupied by the Colonos and their ancestors (COLOMBO, 2007). However, it is possible to see that, since this study, those involved (interviewees) from both groups have shown a relative motivation for aggregation and greater interaction. They suggested that study groups should be developed with interaction between the groups, where topics on the use of medicinal plants in local daily life should be addressed.

CHAPTER 7

CONCLUSION

With regard to the maintenance of ethnopharmacological knowledge, the MST group has been able to do this effectively, while the settler group may be experiencing cultural erosion.

Both groups have extensive ethnopharmacological knowledge, with the MST standing out for citing significantly more. Most of the participants use and trust medicinal plants, and use them on the basis of knowledge obtained from family members.

There was no specific extraction of native species in relation to local medicinal use, with exotic species being used more frequently.

Most of the species inventoried are cultivated, which shows that the local native flora is less vulnerable. However, in order to avoid future problems related to conservation, it is essential that environmental education initiatives are carried out in the area.

The species used in both groups meet local health demands, but it was only possible to support these uses in accordance with ANVISA's RDC for 20 of them. In this way, this work subsidises the choice of plants to be prioritised in the community medicinal garden that is currently being set up.

FINAL CONSIDERATIONS

As reported by the participants in the MST and Colonos groups in the interviews, there is concern about the use and sustainability of natural resources, as seen in the item related to carrying out work aimed at preserving nature.

It is recommended that environmental education projects be developed with an emphasis on methods of sustainable collection and cultivation of medicinal plants.

The communities in question were motivated by this work, which demonstrates the feasibility of developing strategies based on this study that could give greater visibility to the groups studied, which are still neglected by government agencies.

There is a need for aggregation for better development among the groups, and the medicinal garden proved to be a promising factor, as it was accepted by all those interviewed.

BIBLIOGRAPHICAL REFERENCES

AB'SABER, A. N. 2003. The domains of nature in Brazil: Landscape potential. Ateliê editorial.

AGBOR, G.A.; AKINFIRESOYE, L.; SORTINO, J.; JOHNSON, R.; VINSON, J.A.2012. Piper species protect cardiac, hepatic and renal antioxidant status of atherogenic diet fed hamsters. Food Chemistry.134, p. 1354-1359.

NATIONAL HEALTH SURVEILLANCE AGENCY. Resolution - RDC n.10, of 9 March 2010. Provides for the notification of plant drugs to the National Health Surveillance Agency (ANVISA) and other measures.

ALAM, R.; RAHMAN, A. B. ; KADIR, M. M. F.; HAQUE , A.; ALVI, M. R.U.-H. R. 2012. Evaluation of antidiabetic phytochemicals in Syzygium cumini (L.) Skeels (Family: Myrtaceae). Journal of Applied Pharmaceutical Science. 2 (10): 094-098.

ALBERTI, T.B.; MARCON, R,.; BICCA, M.A.; RAPOSO, N.R.; CALIXTO, J.B.; DUTRA, R.C. 2014. Essential oil from Pterodon emarginatus seeds ameliorates experimental autoimmune encephalomyelitis by modulating Th1/Treg cell balance. Journal of Ethnopharmacology. 8:155(1): 485-94.

ALBUQUERQUE, U. P.; ANDRADE, L. H. C. 2002. Traditional botanical knowledge and conservation in a caatinga area in the state of Pernambuco, Northeast Brazil. Acta Botânica Brasílica. 16(3): 273-285.

ALBUQUERQUE, U. P.; LUCENA, R. F. P. 2004. Methods and techniques for data collection. In Methods and Techniques in Ethnobotanical Research. (Ed.). Recife: Livro Rápido/NUPEEA. pp. 39-159.

ALBUQUERQUE, U. P. 2005. Introduction to ethnobotany. 2ª Ed. Rio de Janeiro: Interciência. 80 p.

ALBUQUERQUE, U. P.; HANAZAKI, N. 2006. Ethnodirected research in the discovery of new drugs of medical and pharmaceutical interest: Fragility and prospects. Revista Brasileira de Farmacognosia. 16: 678-689.

ALEIXO, J. 1995. Brazilian Flower Essences. São Paulo: Ground. 185 p.

ALEXÍADES, M. 1996. Selected guidelines for ethnobotanical research: a field manual. New York. The New York Botanical Garden. 306 p.

ALMEIDA, E. R. 1993. Medicinal plants: popular and scientific knowledge. São Paulo: HEMUS. 341 p.

ALMEIDA, L. F. R.; DELACHIAVE, M. E. A.; MARQUES, M. O. M. 2005. Composition of the essential oil of rubim (Leonurus sibiricus L. - Lamiaceae). Revista Brasileira de Plantas Medicinais. 1(8): 35 - 38.

ALMEIDA, M. Z. 2000. Medicinal plants. 1st ed. Salvador: EDUFBA, v. 1, 192p.

ALMEIDA, M. Z. 2003. Medicinal plants. 2nd ed. Salvador: EDUFBA, 150 p.

ALMEIDA, A.P.; RABELO, E. C. 2014. Palavra de Acampado: Experiments in Literary Journalism. Intercom - Brazilian Society for Interdisciplinary Communication Studies XXII Expocom Award - Experimental Communication Research Exhibition.

AMAT, N.; UPUR, H.; BLAZEKOVI, B. 2010. In vivo hepatoprotective activity of the aqueous extract of Artemisia absinthium L: Against chemically and immunologically induced liver injuries in mice. Journal of Ethnopharmacology, 131: 478-484.

AMOROZO, M. C. M.; GÉLY, A. L. 1988. Use of medicinal plants by caboclos from the lower Amazon, Barcarena, PA, Brazil. Boletim do Museu Paraense Emílio Goeldi, Série Botânica v. 4, pp. 47-131.

AMOROZO, M. C.M. 1996. The ethnobotanical approach in medicinal plant research. In: DI STASI, L.C. (Org.) Medicinal Plants: Art and Science, an interdisciplinary study guide. São Paulo: EDUSP. pp. 47-68.

AMOROZO, M. C. M. 2007. Building sustainability: biodiversity in agricultural landscapes and the contribution of ethnobiology. In: Albuquerque, U.P.; Alves, A.G.C.; Araújo, T.A.S. (Org.). People and landscapes: ethnobiology, ethnoecology and biodiversity in Brazil. 1ª ed.Recife: NUPEEA/UFRPE, 2007, v. 1, p. 76-88.

ANTONIO, G.D.; TESSER, C.D.; MORETTI-PIRES, R.O. 2013. Contributions of medicinal plants to care and health promotion in primary care. Interface comunicação saúde educação.

BALÉE, W.; ERICKSON, C. 2006. Time and Complexity in Historical Ecology: Studies in the Neotropical Lowlands. Columbia University Press, New York.

BENNETT, B.C.; PRANCE, G.T. 2000. Introduced plants in the indigenous pharmacopoeia of Northern South America. Economic Botany. 54(1): 90-102.

BOEING, T.; SILVA, L. M.; SOMENSI, L. B.; CURY, B. J.; MICHELS COSTA, A. P.; PETREANU, M.; NIERO, R.; ANDRADE, S. F.; Antiulcer mechanisms of Vernonia condensata Baker: A medicinal plant used in the treatment of gastritis and gastric ulcer. Journal of Ethnopharmacology. 184, p. 196-207, 2016.

BRAZIL 2002 - MINISTRY OF THE ENVIRONMENT. Invasive Alien Species

Ministry of the Environment. Available at: <http://www.mma.gov.br/biodiversidade/biosseguranca/especies-exoticas-invasoras>. Accessed on: 02 September 2016.

BRENNEISEN, E. 2013. The MST and rural settlements in western Paraná: encounters and mismatches in the struggle for land. Studies in Society and Agriculture.

BRITO, A. R. M.; BRITO, A. A. S. 1993. Forty years of Brazilian medicinal plant research. Journal of Ethnopharmacology. 39: 53-67.

BRITO, M. R.; SENNA-VALLE, L.2011. Medicinal plants used in the caiçara community of Praia do Sono, Paraty, Rio de Janeiro, Brazil. Acta Bot. Bras. vol.25, n.2, pp.363- 372.

BURILLE, A.; GERHARDT, T. E. 2014. Chronic diseases, chronic problems: encounters and mismatches with health services in the therapeutic itineraries of rural men. Saúde. vol.23, n.2, pp.664-676.

CAKILCIOGLUA, U.; KHATUNB, S.; TURKOGLUC, I., HAYTAD, S. 2011. Ethnopharmacological survey of medicinal plants in Maden (Elazig-Turkey). J. Ethnopharmacol. 137: 469-486.

CÂMARA, J. B. D; CARVALHO, T. C. S. 2002. (Org.) Brazilian Environment Outlook - GEO Brazil. 1. ed. Brasília: Edições IBAMA, v. 1. 449p.

CAMPÊLO, L. M. L.; GONÇALVES, F. C. M.; FEITOSA, C. M.; FREITAS, R.M. 2011. Evaluation of central nervous system effects of Citrus limon essential oil in mice. Revista Brasileira de Farmacognosia. 21: 668-673.

CAPPELLIN, P.; CASTRO, E. G. 1997. Doing, thinking and deciding: the roles of women in rural settlements. Some reflections based on three case studies. Raízes. 15: 113130.

CARVALHO, A. H. O. ; SANTOS, K. N. S. S.; NUNES, N. S. O. ; RESENDE, D. G.; . REIS. Y. P. B.2008. Verification of the antimicrobial activity of wild plant extracts. Revista Eletrônica de Biologia, v. 1, p. 2-7.

CARVALHO, J. S. B.; MARTINS. J. D. L.; MENDONÇA. M. C. S.; LIMA, L. D. 2013. Popular Use of Medicinal Plants in the Community of Várzea, Garanhuns-PE. Revista Biologia e Ciências da Terra. 13(2): 1519-5228.

NATIONAL FLORA CONSERVATION CENTRE-CNCFLORA. Available at <http://cncflora.jbrj.gov.br/portal> Accessed on: 02 September 2016.

CHEREM, A.R.; TRAMONTE, V. L. C. G.; FETT, R.; VAN DOKKUM, W. 2007. Effect of aubergine peel (Solanum melongena) on plasma concentrations of triglycerides, total cholesterol and lipid fractions in hyperlipidaemic guinea pigs (Cavia porcellus). Rev Bras Pl Med; 9: 51-60.

INTERNATIONAL STATISTICAL CLASSIFICATION OF DISEASES AND HEALTH-RELATED PROBLEMS (CID 10). Available at: <http://www.cid10.com.br/> Accessed on 10 June 2015.

COELI, C. M.; FERREIRA, L. G. F. D.; DE MIRANDA DRBAL, M.; VERAS, R. P., DE CAMARGO JR, K. R.; CASCÃO, Â. M. 2002. Mortality in the Elderly due to Diabetes Mellitus as an underlying and associated cause. Revista de Saúde Pública, 36(2), 135-140.

COLOMBO, A. V. 2002. Fazenda da Fortaleza de Sant'Anna - Historical summary. CHICO BOTICARIO Magazine, Rio Novo, p. 07 - 09.

COLOMBO, A. V.; BARBOSA, C. H. R. 2007. History and Heritage at the Sant Anna Fortress Farm. In: SILVA, W. D. (Org.). Historical and Cultural Aspects of the Municipality of Goianá. 1ed. COMPAC, 2007, v. 01, p. 30-40.

COLOMBO, A. V.; CORRÊA, Â. A. 2014. Babylonian caves: narratives and interventions: pre-colonial funerary remains in the Juiz de Fora micro-region. Cadernos do LEPARQ, v. XI, p. 194-207.

CONDE, B.E. 2012. Ethnopharmacology around the Botanical Garden of the Federal University of Juiz de Fora as a subsidy for the implementation of a community medicinal garden. Master's dissertation in Ecology, Federal University of Juiz de Fora.

CONDE, B.E.; ROGÉRIO, I.T.S.; SIQUEIRA, A.M.; FERREIRA, M.Q.; CHEDIER, L. M.; PIMENTA, D.S. 2014. Ethnopharmacology in the vicinity of the Botanical Garden of the Federal University of Juiz de Fora, Brazil. Ethnobotany Research and Applications. 12: 91111.

CONDE, B. E.; PIMENTA, D. A. 2015. Traditional knowledge vs. modernity: facing the scarcity of natural resources on the planet. In: Tropixel: the arts and sciences. (Ed.) BRUNET, K.; RENNÓ, R. (Org.). Salvador:

EDUFBA.

CONDE, B.E. 2016. Local ecological knowledge and its impact on the conservation of botanical biodiversity for three quilombola communities living in an Atlantic Forest context. Doctoral thesis.

COMPARATO, B. K. 2001. The political action of the MST. São Paulo em Perspectiva (Impresso), São Paulo, v. 15, n.4, p. 105-118.

COMISSÃO PASTORAL DA TERRA (CPT)- Available at <http: www.cptnacional.org.br> Accessed on 06 January 2016.

COSTA, C.A.; KOHN, D.O.; LIMA, V.M; GARGANO A.C.; FLÓRIO, J.C.; COSTA M. 2011. The GABA ergic system contributes to the anxiolytic-like effect of essential oil from *Cymbopogon citratus* (lemongrass) Journal of Ethnopharmacology, 137: 828-836.

COTTON, C.M. 1996. Ethnobotany: principles and applications. Chichester, John Wiley & Sons. 423 p.

CUNHA, M. C.; ALMEIDA, M. W. B. 2001. Traditional populations and environmental conservation. In: CAPOBIANCO, J. P. R.; VERÍSSIMO, A.; MOREIRA, A.; SANTOS, In: PINTO, L. P. 2001. Biodiversity in the Brazilian Amazon: assessment and priority actions for conservation, sustainable use and benefit sharing. São Paulo, Estação Liberdade: Instituto Socioambiental.

D'AQUINO, T. 1998. Settlement as a New Form of Rural Life: Space and Time in the Rural Settlement of Fazenda Reunidas - SP. Raízes 15: 47-61.

DAMASCENA, N. P.; SOUZA, M. T.; ALMEIDA, A. F.; CUNHA, R. S.; CURVELLO, R. L.; LIMA, A.C.; ALMEIDA, E. C.; SANTOS, C. C.; DIAS, A.S . ; PAIXÃO, M. S.; SOUZA, L. M.; QUINTANS JÚNIOR, L. J. ; ESTEVAM, C. S. ; ARAUJO, B. S. ; 2014. Antioxidant and orofacial anti-nociceptive activities of the stem bark aqueous extract of Anadenanthera colubrina (Velloso) Brenan (Fabaceae). Nat. Prod. Res. 28, 753-756.

DE-PARIS, F.; PETRY,R. D.; REGINATTO, F.H.; GOSMANN, G.; QUEVEDO,J.; SALGUEIRO,J. B.; KAPCZINSKI, F.; ORTEGA, G. G.; SCHENKEL,E. P. 2002. Pharmacochemical study of aqueous extracts of Passiflora alata Dryander and Passiflora edulis Sims. Acta Farmaceutica Bonaerense. 21 (1): 5-8.

DIAS, J. E. A. 2002. The importance of the use of medicinal plants in peripheral communities and their production through urban agriculture. Acta Horticulturae. 569: 79-85.

DIEGUES, A. C. S. (org.). 2000. Ethnoconservation: new directions for nature conservation. São Paulo: Hucitec. 290 p.

DI STASI, L. C.; OLIVEIRA, G. P.; CARVALHAES, M. A.; QUEIROZ-JUNIOR, M.; TIEN, O. S.; KAKINAMI, S. H.; REIS, M. S. 2002. Medicinal plants popularly used in the Brazilian Tropical Atlantic Forest. Fitoterapia. 73: 69-91.

DHULEY, J.N. 1999. Anti-oxidant effects of cinnamon (Cinnamomum verum) bark and greater cardamom (Amomum subulatum) seeds in rats fed high fat diet. Indian Journal of Experimental Biology. 37(3): 238-42.

DRUMMOND, G.M.; MARTINS, C.S.; MACHADO, A.B.M.; SEBAIO, F.A.; ANTONINI, Y. 2005. Biodiversity in Minas Gerais: an atlas for its conservation. Biodiversitas Foundation, Belo Horizonte-MG.

DZEREFOS, C.M.; WITKOWSKI, E.T.F. 2001. Density and potential utilisation of medicinal grassland plants from Abe Bailey Nature Reserve, South Africa. Biodiversity and Conservation 10: 1875-1896.

ERHAN, M. K.; BOLUKBASI, S. C.; URUSAN, H. 2012.Biological activities of pennyroyal (Mentha pulegium L.) in broilers. Livestock Science, v. 146, p. 189-192.

EL-SALAM, M. ABD.; MEKKYA, H.; EL-NAGGARB, E. M. B.; GHAREEBC, D. EL- DEMELLAWYD, M..; EL-FIKYA, F. 2015. Hepatoprotective properties and biotransformation of berberine and berberrubine by cell suspension cultures of Dodonaea viscosa and Ocimum basilicum. South African Journal of Botany. 97:191-195.

ELIZABETH, A. A.; JOSEPHINE, G.; MUTHIAH, N.S.; MUNIAPPAN, M. 2014. Evaluation of Analgesic and Anti-Inflammatory Effect of Foeniculum vulgare. Research Journal of Pharmaceutical, Biological and Chemical Sciences 5(2): 658-668.

ETKIN, N. L.; E. ELISABETSKY. 2005. Seeking a transdisciplinary and culturally germane science: The future of ethnopharmacology." Journal of ethnopharmacology. 100(1): 23-26.

FALEIROS, R. N. 2007. Frontiers of coffee: farmers and settlers in the interior of São Paulo (19171937). UNICAMP-SP. COLOMBO, A. V. Fazenda da Fortaleza de Sant'Anna - Historical summary. CHICO BOTICARIO Magazine, Rio Novo, p. 07 - 09.

FARIAS, L. B. P. 2016. The voice of witches! The speech of MST women in the health sector of the Zona da Mata Mineira. In: FERRANTE V. L. S. B. Retratos de Assentamentos, vol.19 n.1. The Rural Research and Documentation Centre (Nupedor) linked to the Postgraduate Programme in Territorial Development and the Environment - UNIARA.

FERNANDES, B. M. A. 2000. The formation of the MST in Brazil. Digital Library of the Brazilian Agrarian Question. Available at http://www.reformaagrariaemdados.org.br/biblioteca/livros. Accessed on 03 July 2016.

BIODIVERSITAS FOUNDATION. 2009. Instituto de Pesquisas Ecológicas, Secretaria do Meio Ambiente do Estado de Minas Gerais & Instituto de Florestas.

FLORA BRASILIENSIS. 2014. Selection of species native to Brazil. Available at<:http://florabrasilensis,cria.org.br>Accessed on 06 January 2016.

GALEANO, G. 2000. Forest use at the pacific coast of Choco, Colombia: a quantitative approach. Economic Botany. 54: 258-376.

GERHARDT, T. E.; BURILLE, A. 2014. Chronic diseases, chronic problems: encounters and mismatches with health services in the therapeutic itineraries of rural men. Saúde Soc. 23(2): 664-676.

GONZALEX-LIMA, F.; VALEDON, A.; STIEHIL, W.L. 1986. Depressant pharmacological effects of an isolated component of lettuce, Lactuca sativa. Pharmaceutical Biology. 24 : 154-66.

GONZÁLEZ-TRUJANO, M. E.; PELLICER, F.; MENA, P.; MORENO, D. A.; GARCÍA- VIGUERA, C. 2015. Antinociceptive and anti-inflammatory activities of a pomegranate (Punica granatum L.) extract rich in ellagitannins. International Journal of Food Sciences and Nutrition. 66 (4): 395-399.

GOTTLIEB, O. R.; KAPLAN, M. A.C.; BORIN, M. R. R. B. 1996. Biodiversity. A chemical-biological approach. Rio de Janeiro: UFRJ, 268p.

GRADINARIU, V.; CIOANCA, O.; HRITCU, L.; TRIFAN, A.; GILLE, E.; HANCIANU, H. 2015.

Comparative efficacy of Ocimum sanctum L. and Ocimum basilicum L. essential oils against amyloid beta (1-42)-induced anxiety and depression in laboratory rats. Phytochemistry Reviews. 14 (4): 567-575.

GUATURA, I.N.; CORREA, E. COSTA, J. P. O. 1996. A questão fundiária: Roteiro para a solução dos problemas fundiários nas áreas protegidas da Mata Atlântica: Roteiro para a conservação de sua biodiversidade. São Paulo: National Council of the Atlantic Forest Biosphere Reserve, 47 p.(Cadernos da Reserva da Biosfera,1).

GUIMARÃES, E. S. 2009. Autonomous slave economy in the large coffee plantations of the southeast. Latin America in Economic History. 32: 155-187.

HALAL, I. S. F.; BERTONI, A. M.; CIACOMET, C.; SEIBEL, C. E.; LAHUDE, F. M.; MAGALHÃES, G. A. 1994. Evaluation of the quality of primary health care in an urban locality in southern Brazil. Revista de Saúde Pública. 28(2): 131-136.

HAN, Y.A.; SONG, C.W.; KOH, W.S.; YON, G.H.; KIM, Y.S.; RYU, S.Y.; KWON, H.J.; LEE, K.H. 2013. Anti-inflammatory effects of Zingiber officinale Roscoe constituent 12 dehydrogingerdione on lipopolysaccharide-stimulated RAW 264.7 cells. Phytotherapy Research. 27 (8): 1200-5.

HANAZAKI, N.; TAMASHIRO, J. Y.; LEITÃO-FILHO, H. F.; BEGOSSI, A. 2000. Diversity of plant uses in two Caiçara communities from the Atlantic Forest coast, Brazil. Biodiversity and Conservation. 9: 597-615.

HANAZAKI, N.; SOUZA V.C; RODRIGUES, R. R. 2006. Ethnobotany of rural people from the boundaries of Carlos Botelho State Park, São Paulo State, Brazil. Acta Botânica Brasílica. 20: 899-909.

HERNÁNDEZ, A. I. G.; BARREIRO, M. L.; RODRÍGUEZ, Z. M.; ALVAREZ, G.B.; MORÓN RODRÍGUEZ, F. J.; RODRÍGUEZ E. B.; AMADOR, M. C. V.; HORMAZA, I. M.; LA LUZ, C. L. L. A.; GONZÁLEZ,A. D.; ALVAREZ, G. B.; RODRÍGUEZt, F. J. M. 2014. Preclinical validation of the peripheral and central analgesic activity of decoction of fresh leaves of Persea americana Mill. (avocado) and Musa x paradisiaca L. (banana)]. Revista Cubana de Plantas Medicinales 19 (3): 225-234.

HEINRICH, M. BARNES, J.; GIBBONS, S.; WILLIAMSON, E.M. 2004. Fundamentals of pharmacognosy and phytotherapy, London: Churchill Livingstone.

HOSSAIN, H.; KARMAKAR, U.K.; BISWAS, S.K.; SHAHID-UD-DAULA, A.F.; JAHAN, I.A.; ADNAN, T.; CHOWDHURY, A. 2013. Antinociceptive and antioxidant potential of the crude ethanol extract of the leaves of Ageratum conyzoides grown in Bangladesh. Pharmaceutical Biology. 51(7): 893-8.

HUSSAIN, A.; SHARMA, K. C. S. ; SHAH K. ; HAQUE, S. 2015. Aloe vera Inhibits Proliferation of Human Breast and Cervical Cancer Cells and Acts Synergistically with Cisplatin. Asian Pacific Journal of Cancer Prevention. 16 (7): 2939-2946.

HUSSAINI, R.A.; MAHASNEH, A.M. 2009. Antimicrobial and antiquorum sensing activity of different parts of Laurus nobilis L. extracts Phytotherapy Research. 17 (7): 733-736.

IGNACIO, S. R. N.; FERREIRA, J. L. P.; ALMEIDA, M. B.; KUBELKA, C. F., 2001: Nitric oxide production by murine peritoneal macrophages in vitro and in vivo treated with *Phyllanthus tenellus* extracts. Journal of Ethnopharmacology 74(2): 181-187.

INDHUMATHI.T.; KANNIKAPARAMESWARI, K. 2012. Haematopoietic study of the methanolic root extract of Beta vulgaris on albino rats-an in vivo study. International Journal of Pharma and Bio Sciences 3(4): 1005 - 1015.

INCRA - NATIONAL INSTITUTE FOR COLONISATION AND AGRARIAN REFORM MINISTRY OF AGRARIAN DEVELOPMENT. 2016. Available at <http://www.incra.gov.br/assentamentcessado> on 04 January 2016.

ISLAM, M. A.; AHMED, F.; DAS, A.K.; BACHAR, S.C. 2005. Analgesic and anti- inflammatory activity of Leonurus sibiricus. Phytotherapy, 76: 359-362.

IUCN - INTERNATIONAL UNION FOR CONSERVATION OF NATURE (IUCN). 2016. 'Red List of Threatened Plants'. Available at <http://www.iucnredlist.org> Accessed on 06 January 2016.

JIVAD, N.; BAHMANI, M. 2016. A review of important medicinal plants native to Iran effective on recovery from peptic ulcer. Der Pharmacia Lettre, v 8, p-347-352.

JORGE, T.C.; LENARTOVICZ, V. ; ANDRADE, M.W .; BONAFIN, T. .;GIORDANI, M.A. ; BUENO, N.B .;SCHNEIDER, D.S.L.G. 2009. Pediculicidal activity of hydroethanolic extracts of Ruta graveolens, Melia azedarach and Sambucus australis. Lat. Am. *J. Pharm,* 28(3), pp.457-459.

KARIMA, S.; FARIDA, S.; MIHOUB, Z.M. 2015. Antioxidant and antimicrobial activities of Plantago major. International Journal of Pharmacy and Pharmaceutical Sciences 7 (5): 58-64

KHARE, P.; MISHRA, V.K.; ARUN, K.; BAIS, N.; SINGH, R. 2014. Study on in vitro anti lithiatic activity of Phyllanthus niruri linn. leaves by homogenous precipitation and turbiditory method. International Journal of Pharmacy and Pharmaceutical Sciences. 6 (4): 124-127.

KHODABAKHSH, P.; SHAFAROODI, H.; ASGARPANAH, J. 2015. Analgesic and anti- inflammatory activities of Citrus aurantium L. blossoms essential oil (neroli): involvement of the nitric oxide/cyclic-guanosine monophosphate pathway. Journal of Natural Medicines. 69(3):324-331.

KURTZ, B. C.; ARAÚJO, D. S. D. 2000. Floristic composition and structure of the tree component of a stretch of Atlantic Forest in the Paraíso State Ecological Station, Cachoeiras de Macacu, Rio de Janeiro, Brazil. Rodriguésia. 51:69-12.

LEAL, L. K. A. M.; FERREIRA, A. A. G.; BEZERRA, G. A.; MATOS, F. J. A.; VIANA, G. S. B. 2000. Antinociceptive, anti-inflammatory and bronchodilator activities of Brazilian medicinal plants containing coumarin: a comparative study. Journal of Ethnopharmacology. 70: 151-159.

LEE, J.; CHOI, J.; KIM, S. 2015. Effective suppression of pro-inflammatory molecules by DHCA via IKK-NF-κB pathway, in vitro and in vivo British Journal of Pharmacology, v 172, P. 3353-3369.

LEE, J.B.; MIYAKE, S. ; UMETSU, R. ; HAYASHI, K . ; CHIJIMATSU, T. ; HAYASHI, T. 2012. Anti-influenza A virus effects of fructan from Welsh onion (Allium fistulosum L.). Food Chemistry. 134 (4): 2164-8.

LEE, J.G.A.; JIN, J.H.A.;LIM, H.T.B.; CHOI, H.D.C.;KIM, H.P. 2009.A"Inhibition of Experimental Gastric Ulcer by Potato Tubers and the Starch." Natural Product Sciences 15.3 : 134-138.

LEITÃO-FILHO, H. de F. 1987. Considerations on the floristics of tropical and subtropical forests in Brazil. IPEF, v. 35, p. 41-46

LI, H.; DENG, Z.; LIU, R.; LOEWEN, S.; TSAO, R. 2014. Bioaccessibility, in vitro antioxidant activities and in vivo anti-inflammatory activities of a purple tomato (Solanum lycopersicum L.) Food Chemistry. 159: 353-360.

LIMA, R. X. ; SILVA, S. M.; KUNIYOSHI, Y. S.; SILVA, L. B. 2000. Ethnobiology of continental communities in the Guaraqueçaba Environmental Protection Area, Paraná, Brazil. Etnoecológica. 4(6): 33-55.

LIZ, R.; VIGIL, S.V.G.; GOULART, S.; MORITZ, M.I.G.; SCHENKEL, E.P.; FRODE. 2008. T.S. The anti-inflammatory modulatory role of Solidago chilensis Meyen in the murine model of air pouch. Journal of Pharmacy and Pharmacology. v. 60, pp. 515-521.

LOPES, L.C.; CARVALHO, J.E; KAKIMORE, M.; VENDRAMINI-COSTA, D.B.; MEDEIROS, M.A.; SPINDOLA, H.M.; ÁVILA-ROMÁN, J.; LOURENÇO, A.M.; MOTILVA, V. 2014. Pharmacological characterisation of Solanum cernuum Vell.: 31- norcycloartanones with analgesic and anti-inflammatory properties. Inflammopharmacology, 22(3), pp.179-185.

LOMBARDO-EARL, G.; ROMAN-RAMOS R.; ZAMILPA, A.; HERRERA-RUIZ M.; ROSAS-SALGADO, G.; TORTORIELLO, G.2014. Extracts and Fractions from Edible Roots of Sechium edule (Jacq.) Sw. with Antihypertensive Activity. Evidence-Based Comp Alt Med. 1-9.

LUCENA, R.F.P.; LUCENA, C.M.; ARAÚJO, E.L.; ALVES, A.G.C.; ALBUQUERQUE, U. P. 2013. Conservation priorities of useful plants from different techniques of collection and analysis of ethnobotanical data. Annals of the Brazilian Academy of Sciences. 85 (1): 169-186.

MACDONALD, D.; VANCREY, K.; HARRISON, P.; RANGACHARI ,P.; ROSENFELD, J. ; WARREN, C.; SORGER, G. 2004. Ascaridole-less infusions of *Chenopodium ambrosioides* contain a nematocide(s) that is(are) not toxic to mammalian smooth muscle. Journal of Ethnopharmacology. 92(2-3): 215-21.

MACHADO, D.G.; BETTIO, L.E.B.; CUNHA, M.P.; CAPRA, J.C.; DALMARCO, J.B.; PIZZOLATTI, M.G.; RODRIGUES, A.L.S. 2009. Antidepressant-like effect of the extract of Rosmarinus officinalis in mice: Involvement of the monoaminergic system. Progress in Neuro-Psychopharmacology & Biological Psychiatry. 33, p. 642-650.

MAIOLI-AZEVEDO, V.; FONSECA-KRUEL, V. S. 2007. Medicinal and ritualistic plants sold in open-air markets in the city of Rio de Janeiro, RJ, Brazil: a case study in the North and South zones. Acta Botânica Brasílica. 21: 263-275.

MALIPATIL, N.B.A.; MANJUNATH, S.B.; SHRUTHI, D.P.C. 2015. Evaluation of effect of aqueous extract of Zingiber Officinale Roscoe (Ginger) on acute and chronic inflammation in adult albino rats (Article). Asian Journal of Pharmaceutical and Clinical Research 8(2): 113-116

MARORYI, A. 2011. An.ethnobotanical survey of medicinal plants used by the people in Nhema communal area, Zimbabwe. Journal of Ethnopharmacology. 136: 347-354.

MARTINS, E. R.; CASTRO, D. M. DE; CASTELLANI, D. C.; DIAS, J. E. 2000. Medicinal Plants. Viçosa: Editora da UFV: Federal University of Viçosa, 220 p.

MAZUMDAR, S.; AKTER, R.; TALUKDER, D. 2015. Antidiabetic and antidiarrhoeal effects on ethanolic extract of Psidium guajava (L.) Bat. leaves in Wister rats. Asian Pacific Journal of Tropical Biomedicine. 5(1): 10-14.

MELO, G.O.; MALVAR, D. C.; VANDERLINDE, F.A.; ROCHA, F.F.; PIRES, P.A.; COSTA, E.A.; MATOS, L.G.; KAISER, C.R.; COSTA, S.S.2009. Antinociceptive and anti- inflammatory kaempferol glycosides from Sedum dendroideum. Journal Ethnopharmacol. 15;124(2):228-32.

MELLO, V.; PRATA, M.C .; SILVA, M.R.; DAEMON, E.; SILVA L. S.; GUIMARÃES, F.; DEL G .; MENDONÇA, A.E.; FOLLY, E .; VILELA, F.M .; AMARAL, L.H .; CABRAL, L.M.; AMARAL, M. P . 2014. Acaricidal properties of the formulations based on essential oils from Cymbopogon winterianus and Syzygium aromaticum plants. Parasitology Research. 113 (12): 4431-7.

MENDIETA, M. D. C.; DE SOUZA, A. D. Z.; VARGAS, N. R. C., PIRIZ, M. A.; ECHEVARRÍA-GUANILO, M. E.; HECK, R. M. 2014. Transmission of knowledge about medicinal plants in the family context: integrative review. Revista de Enfermagem da UFPE, Recife. 8(10): 3516-3524.

MICHEL, A.F.; MELO, M.M.; CAMPOS, P.P.; OLIVEIRA, M.S.; OLIVEIRA, F.A.; CASSALI, G.D.; FERRAZ, V.P.; COTA, B.B.; ANDRADE, S.P. ; SOUZA-FAGUNDES, E.M . 2015. Evaluation of anti-inflammatory, antiangiogenic and antiproliferative activities of Arrabidaea chica crude extracts. Journal of Ethnopharmacology. 165: 29-38.

MINISTRY OF THE ENVIRONMENT SECRETARIAT OF BIODIVERSITY AND FORESTS

DEPARTMENT OF BIODIVERSITY CONSERVATION. 2008. Red Book of Endangered Brazilian Fauna. Biodiversity. Brasília, DF.

MIRANDA, G.S.; SANTANA, G.S.; MACHADO B.B.; COELHO, F.P.; CARVALHO, C.A. 2013. In vitro antibacterial activity of four plant species in different alcoholic strengths. Revista brasileira de plantas medicinais v.15 no.1 Botucatu .

MOHAMMED, A.; TANKO, Y.; OKASHA, M. A..; MAGAJI, R. A. ; YARO, A. H. 2007. Effects of aqueous leaves extract of Ocimum gratissimum on blood glucose levels of streptozocininduced diabetic wistar rats. African Journal of Biotechnology, 6(18).

MONTERROSAS-BRISSON,N.;OCAMPO, M. L. A.; JIMÉNEZ-FERRER E.; JIMÉNEZ- APARICIO,A. R.; ZAMILPA , A.; GONZALEZ-CORTAZAR,M.; TORTORIELLO,J.; HERRERA-RUIZ,M. 2013. Anti-inflammatory activity of different agave plants and the compound cantalasaponin- Molecules.;18 (7):8136-8146.

MOURÃO, R. H.; SANTOS, F. O.; FRANZOTTI, E. M.; MORENO, M. P. 1999. Anti-inflammatory activity and acute toxicity (LD50) of the juice of Kalanchoe

brasiliensis (Comb.) leaves picked before and during blooming.Phytotherapy Research, v. 13, p. 352-354.

MST-LANDLESS RURAL LABOUR MOVEMENT. 2016. <http://www.mst.org.br>Accessed on 06 January 2016.

MUSS, C.; MOSGOELLER, W.; ENDLER, T. 2013. Papaya preparation (Caricol®) in digestive disorders. Neuroendocrinology Letters. 34(1): 38-46.

MYERS, N.; MITTERMEIER, R. A.; MITTERMEIER, C.G.; FONSECA, G.A.B.; KENT, J. 2000. Biodiversity hotspots for conservation priorities. Nature, 403(6772): 853-858.

NAGAPPAN, P.; GOMATHINAYAGAM, S. 2014. Study of Mosquito Larvicidal Effects of Momordica charantia (Bitter Gourd) Extracts as Nanopowder. International Journal of ChemTech Research 6(9): 4052-4054.

NARANJO, J.P.; CUBILES, M.A.M.; SALVADÓ, A.C.; CAMPOS, C.G. 2006. Actividad antiparasitaria de una decocción de *Mentha piperita* Linn, in Revista Cubana de Medicina Militar. 35(3): 1-4.

NASCIMENTO, G.E.D.; BAGGIO, C.H.; WERNER, M.F.D.P.; IACOMINI, M.; CORDEIRO, L.M. 2016. Arabinoxylan from Mucilage of Tomatoes (Solanum lycopersicum L.): Structure and Antinociceptive Effect in Mouse Models. Journal of agricultural and food chemistry, 64(6), pp.1239-1244.

NDOYE,F. F.M.; TCHINANG,T.F.; NYEGUE, A.M.; ABDOU, J.P.; YAYA, A.J.;TCHINDA, A.T.; ESSAME, J.L.; ETOA, F.X. 2016. Chemical composition, in vitro antioxidant and anti-inflammatory properties of essential oils of four dietary and medicinal plants from Cameroon. Evidence-Based Complementary and Alternative Medicine. 7;16:117. doi: 10.1186/s12906-016-1096-y.

NEGRI, L. ; MATTEI, R.; MENDES ,F.R .2014 Antinociceptive activity of the HPLC- and MS-standardised hydroethanolic extract of Pterodon emarginatus Vogel leaves. Phytomedicine. 21(8-9):1062-9.

NEVES, D. P. 1999. Rural settlement: confluence of forms of social insertion. Revista Estudos Sociedade e Agricultura, Rio de Janeiro, v. 13, p. 5-28.

OBOH, G.; ADEFEGHA, S.A.; ADEMOSUN, A. O.; UNU, D. 2010. Effects of hot water treatment on the phenolic phytochemicals and antioxidant activities of lemon grass *(Cymbopogon citratus)* ejeafche, 9 (3): 503-513.

OKUMA, C.H.; ANDRADE, T.A; CAETANO, G.F. ; FINCI, L.I. ; MACIEL, N.R. ; TOPAN J.F. ; CEFALI, L.C. ; POLIZELLO, A.C. ; CARLO, T.; ROGERIO. A.P. ; SPADARO A.C. ; ISAAC, V.L. ; FRADE, M.A. ; ROCHA-FILHO, P.A. 2015. Development of lamellar gel phase emulsion containing marigold oil (Calendula officinalis) as a potential modern wound dressing. European Journal of Pharmaceutical Sciences. 71:62-72.

OLIVEIRA, E. R; MENINI NETO, L. 2012. Ethnobotanical survey of medicinal plants used by residents of the village of Manejo, Lima Duarte - MG. Revista Brasileira de Plantas Medicinais, vol.14, n.2, pp.311-320.

ABREU, O. A.; SÁNCHEZ, I.; PINO, JORGE.; BARRETO, G.2015. Antimicrobial activity of piper aduncum subsp ossanum essential oil. International Journal of Phytomedicine, v.7,n.2.

PARISE-FILHO, R.; PASTRELLO, M.; PEREIRA CAMERLINGO, C.E.; SILVA, G.J.; AGOSTINHO, L.A.; DE SOUZA, T.; MOTTER MAGRI, F.M.;RIBEIRO, R.R.; BRANDT, C.A.; POLLI, M.C. 2011. The anti-inflammatory activity of dillapiole and some semisynthetic analogues. Pharmaceutical biology, 49(11), pp.1173-1179.

PASSALACQUA, N.G.; GUARRERA, P.M.; DE FINE G. 2007. Contribution to the knowledge of the folk plant medicine in Calabria region (Southern Italy). Fitoterapia. 2007 Jan; 78(1): 52-68.

PASTORE, E. 2005. Gender relations in ecological agriculture. Journal of Economic Theory and Evidence, University of Passo Fundo. FEAC Research and Extension Centre.

PATEL, S.; SHARMA, V.; CHAUHAN, N.S. ;THAKUR, M; .DIXIT ,V.K. 2015. Hair Growth: Focus on Herbal Therapeutic Agent. journal of ethnopharmacology Curr Drug Discov Technol. 12(1):21-42.

PAULA-FREIRE, L. I. G.; ANDERSEN, M. L.; MOLSKA, G. R.; KOHN, D. O.; CARLINI, E. L. A. (2013). Evaluation of the antinociceptive activity of Ocimum gratissimum L.(Lamiaceae) essential oil and its isolated active principles in mice. Phytotherapy Research, 27(8), 1220-1224

PAULO, P. T. C.; DINIZ, M. F. F. M.; MEDEIROS, I. A.; MORAIS, L. C. S. L.; ANDRADE, F. B.; SANTOS, H. B. 2009. Phase I clinical toxicological trials of a compound herbal medicine (Schinus terebinthifolius Raddi, Plectranthus amboinicus Lour and Eucalyptus globulus Labill). Revista brasileira de farmacognosia. vol.19 no.1. João Pessoa- PB.

PAVAN-FRUEHAUF. S. 2000. Medicinal plants of the Atlantic Forest: Sustainable management and sampling. Annablume-Fapesp. São Paulo.

PEIXOTO, A. L.; ROSA, M. M. T. ; SILVA, I. M.. 2002. Characterisation of the Atlantic Forest. In: Sylvestre, L.S. & Rosa, M.M.T.. (Methodological manual for botanical studies in the Atlantic Rainforest. 1ª ed. Seropédica, RJ: Editora Universidade Rural, v. 1, p. 9-23.

PERANDIN, DIEGO.; MAIOLI, M. A. ; SANTOS, P. R. S.; PEREIRA, F. T.V.; MINGATTO, F. E. 2015. Protection of iron-induced liver oxidative damage by

aqueous extract of the plant Plectranthus barbatus. Brazilian Journal of Medicinal Plants. 17: 9-17.

PEREIRA, I. G. R. 2002. Prevalence of the use of phytotherapy in patients in the geriatrics programme at the University Hospital of Brasília - HUB. (Master's dissertation in Health Sciences) University of Brasília.

Brasília, DF.

PHILIPS, O.; GENTRY, A. H. 1993. The useful plants of Tambopata, Peru: I. Statistical hypothesis tests with a new quantitative technique. Economic Botany.47, 15-32.

PINTO, E. P. P.; AMOROZO, M. C. M.; FURLA, A.2006. Popular knowledge of medicinal plants in rural communities in the Atlantic Forest - Itacaré, BA, Brazil Acta bot. bras. 20(4): 751-762.

PIRES, J. M.; MENDES, F. R.; NEGRI, G.; DUARTE-ALMEIDA, J.M.; CARLINI, E.A. 2009. Antinociceptive peripheral effect of Achillea millefolium L. and Artemisia vulgaris L.: both plants popularly known by brand names of analgesic drugs. Phytoterapy Research. 23(2): 212-9.

PONGTHANAPISITH, V.; IKUTA, K.; PUTHAVATHANA, P.; LEELAMANIT, W. 2013. Antiviral protein of Momordica charantia L. inhibits different subtypes of Influenza A. Evidence-Based Complementary and Alternative Medicine 729081.

BIODIVERSITY PORTAL OF THE MINISTRY OF THE ENVIRONMENT. Available at < https://portaldabiodiversidade.icmbio.gov.br/instituicoes>Accessed on 29 August 2016 .

PUSHPAN, R.; NISHTESWAR, K.; KUMARI, H. 2012. Ethno medicinal claims of Leonotis nepetifolia (L.) R. Br: A review (Review). International Journal of Research in Ayurveda and Pharmacy. 3(6): 784-785.

QUINTANS JÚNIOR, L.J.; SANTANA, M.T.; MELO, M.S.; SOUSA D.P.; SANTOS I.S.; SIQUEIRA, R.S.; LIMA T.C.; SILVEIRA G.O.; ANTONIOLLI, A.R.; RIBEIRO L.A.; SANTOS, M.R. 2010. Antinociceptive and anti-inflammatory effects of Costus spicatus in experimental animals. Pharmaceutical Biology 48(10):1097-102.

RABELO, M.; SOUZA, E.P.; SOARES, P.M.G.; MIRANDA,,A.V.; MATOS, F.J.A. ; CRIDDLE, D.N. 2003.Antinociceptive properties of the essential oil of Ocimum gratissimum L. (Labiatae) in mice. Brazilian Journal Medical Biological Research, v.36(4) 521-524 (Short Communication).

RADHIKA, B.; BEGUM, N.; SRISAILAM, K.; REDDY, V. M. 2010. Diuretic activity of Bixa orellana linn. leaf extracts. Indiam Journal of Natural Products and Resources 1(3): 353355.

RATHOD, S.; PHIRI, P.; HARRIS, S.; UNDERWOOD, C.; THAGADUR, M.; PADMANABI, U.;

KINGDON, D. 2013. Cognitive behaviour therapy for psychosis can be adapted for minority ethnic groups: a randomised controlled trial.

RATTNER, H. 2009. Environment, health and sustainable development. Ciência saúde coletiva vol.14 n°6 Rio de Janeiro, p.1678-4561.

RIBEIRO, L. M. P. 2004. The Meek Shall Inherit the Earth - An Ethnobotanical Study of a Protestant Rural Area. 01. Edition. São Paulo: Editora Mackenzie, v. 01. 382p.

ROGÉRIO, I. T. S. Ethnobotany in the quilombola community of São Bento, Municipality of Santos Dumont/MG. 2014. Dissertation (Master's Degree in Postgraduate Programme in Ecology - PGECOL) - Federal University of Juiz de Fora, Juiz de Fora.

RODRIGUES, E.; CARLINI, E. L. DE. 2003. Ethnopharmacological survey of a group of quilombolas in Brazil. Arq Bras Fitomed Cient. 1 (92): 80-7.

RODRIGUES, J. F. L.; SANTOS, S. M.; SOUZA, J. H. G. ; LAGO. 2012. The mystery of the 'resin-of-canuaru': A medicine used by caboclosriver-dwellers of the Amazon, Amazonas, Brazil. Journal of Ethnopharmacology , v.144, p. 806-808.

ROSSATO, S.; LEITÃO-FILHO, H. ; BEGOSSI, A. 1999. Ethnobotany of Caiçaras of the Atlantic Forest Coast (Brazil). Economic Botany , v.53, n.4,p.387-395.

SABITHA, V.; RAMACHANDRAN, S.; NAVEEN, K. R.; PANNEERSELVAM, K. 2011. Antidiabetic and antihyperlipidemic potential of Abelmoschus esculentus (L.) Moench. in streptozotocin-induced diabetic rats. Journal of Pharmacy and Bioallied Sciences 3(3): 397-402.

SADEGHI, H.; MOSTAFAZADEH, M.; SADEGHI, H.; NADERIAN, M.; BARMAK ,M.J.; TALEBIANPOOR, M.S.; MEHRABAN, F. 2014. In vivo anti-inflammatory properties of aerial parts of Nasturtium officinale. Pharmaceutical Biology 52(2):169-74.

SANTOS, A. R.; FILHO, V. C.; NIERO, R.; VIANA, A. M.; MORENO, F. N.; CAMPOS, M. M.; YUNES, R. A.; CALIXTO, J. B. 1994. Analgesic eff ects of callus culture extracts from selected species of *Phyllanthus* in mice. J Pharm Pharmacol. 46: 755-759.

SANTOS, J. S.; MARINHO, R. R.; EKUNDI-VALENTIM, E.; RODRIGUES, L.; YAMAMOTO.; M. H.; TEIXEIRA, S. A.; MUSCARA, M. N.; COSTA, S.K.; THOMAZZI, S. M. 2013. Beneficial effects of Anadenanthera colubrina (Vell.) Brenan extract on the inflammatory and nociceptive responses in rodent models. Journal of Ethnopharmacology: 148: 218-222.

SANTOS, M.W.2013. Quilombola festivities: Between tradition and the sacred, shades of African ancestry. Revista HISTEDBR. 50: 286-300.

SANTOS, S. C; MELO, U.S.; LOPES, S. S. S; WELLER, M; KOK, A. F. 2013. Could inbreeding explain the high prevalence of disabilities in populations in the Brazilian Northeast? Ciência saúde coletiva. 18(4): 1141-1150.

SARAIVA, L. F. 2005. Land structure and labour transition in a major coffee centre, Juiz de Fora - 1870 - 1900. Revista Científica da FAMINAS. 2(2): 179-212.

SASAKI, K.; OMRI, A.E.; KONDO, S.; HAN, J.; ISODA, H. 2013. Rosmarinus officinalis polyphenols produce antidepressant like effect through monoaminergic and cholinergic functions modulation. Behavioural Brain Research. 238, p. 86-94.

SAYYAH, M.; SAROUKHANI,G.; L.; PEIROVI, A.; KAMALINEJAD, M . 2003. Analgesic and anti-inflammatory activity of the leaf essential oil of Laurus nobilis Linn. Phytotherapy Research. 17(7): 733-6.

SHAFI,G.; HASAN, T. N.; SYED, N. A.; AL-HAZZANI A. A.; ALSHATWI A. A.; JYOTHI A.; MUNSHI A. 2012. Artemisia absinthium (AA): a novel potential complementary and alternative medicine for breast cancer. Molecular Biology Reports. 39(7): 7373-7379.

SHROTRIYA, S.; ALI, M.S.; SAHA, A.; BACHAR, S.C.; ISLAM, M.S. 2007. Anti- inflammatory and analgesic effects of hedychium coronarium koen. Pak. Journal of Pharmaceutical Sciences 20(1): 42-47.

SILVA, A. G.; MACHADO, E. R.; ALMEIDA, L.M.; NUNES, R.M .; GIESBRECHT, P.C.; COSTA, R. M.; COSTA, H. B.; ROMÃO, W .; KUSTER, R.M . 2015. A Clinical Trial with Brazilian Arnica *(Solidago chilensis* Meyen) Glycolic Extract in the Treatment of Tendonitis of Flexor and Extensor Tendons of Wrist and Hand. Phytother Research. 29(6): 864-9.

SILVA, A. T. R. 2015. Biodiversity conservation between traditional knowledge and science. Science, values and alternatives II, Brasília, vol.29, n.83, pp.233-259.

SILVA, F. D. S.; RAMOS, M. A.; HANAZAKI, N.; ALBUQUERQUE, U. P. D. 2011. Dynamics of traditional knowledge of medicinal plants in a rural community in the Brazilian semi-arid region. Revista Brasileira de Farmacognosia, 21(3), 382-391.

SILVA, N. C. B.; DELFINO-REGIS, A. C. S.; ESQUIBEL, M. A.; SANTOS, J. E. S.; ALMEIDA, M. Z. 2012. Medicinal plants use in Barra II quilombola community-Bahia, Brazil. Bol. Latinoam. Caribe Plant. Med. Aromat. 11: 435-453.

SILVA, N.C.C.; BARBOSA, L.; SEITO, L.N.; FERNANDES JUNIOR, A. 2012. Antimicrobial activity and phytochemical analysis of crude extracts and essential oils from medicinal plants. Natural Product Research: Formerly Natural Product Letters. v. 26.

SILVA, S. M. P.; MORAES, I. F. 2008. Family farming and the National Medicinal Plants and Phytotherapics Programme: How public policy can make this production chain viable. Revista Tecnologia & Inovação Agropecuária, Pindamonhangaba-SP, p.67-76.

SILVA, T. S. C.; SUFFREDINI, I. B.; RICCI, E. L.; FERNANDES, S. R. C.; GONÇALVES, V. JR.; ROMOFF, P.; LAGO, J. H. G.; BERNARDI, M. M. 2015. Antinociceptive and anti- inflammatory effects of Lantana camara L. extract in mice. Brazilian Journal of Medicinal Plants. 17(2): 224-229.

SIQUEIRA, A.M. 2014. Ethnobotany in the quilombola community of São Sebastião da Boa Vista, Municipality of Santos Dumont/MG. 2014. Dissertation (Master's Degree in Postgraduate Programme in Ecology - PGECOL) - Federal University of Juiz de Fora, Juiz de Fora.

SOARES de MOURA, R.; COSTA, S. S.; JANSEN, J. M.; SILVA, C. A.; LOPES, C. S.; BERNARDO-FILHO, M.; NASCIMENTO da SILVA, V.; CRIDDLE, D. N.; PORTELA, B.N.; RUBENICH, L.M.; ARAUJO, R.G.; CARVALHO, L. C. 2002. Bronchodilator activity of Mikania glomerata Sprengel on human bronchi and guinea-pig trachea. Journal of Pharmacy and Pharmacology. 54: 249-256.

SOUZA, S.D.; FRANCA, C.S .; NICULAU, E.S.; COSTA, L.C.; PINTO, J.E.; ALVES, P.B.; MARÇAL, R.M . 2015. Antispasmodic effect of Ocimum selloi essential oil on the guinea-pig ileum. Natural Product Research 29 (22): 2125-8.

SOUZA, S. P.; PEREIRA, L. L. S.; SOUZA, A. A.; SOUZA, R. V.; SANTOS, C. D. 2012. Study of the anti-obesity activity of the methanolic extract of Baccharis trimera (Less.) DC. Revista Brasileira de Farmácia.

(93): 27-32.

SRIVASTAVA, S.; CHANDRA, D . 2013. Pharmacological potentials of Syzygium cumini: a review. Journal of the Science of Food and Agriculture. 93(9): 2084-93.

STANISKI, A.; FLORIANI, N.; STRACHULSKI, J. 2014. Ethnobotanical study of medicinal plants in the Sete Saltos de Baixo faxinal community, Ponta Grossa - PR. Terra Plural. 8(2): 321-340.

STEHMANN, J. R.; FORZZA , R. C.; SALINO, A.; DA COSTA, D. P.; KAMINO, L. H. Y. 2009. Plants of the Atlantic Forest. Rio de Janeiro, Rio de Janeiro Botanical Garden.

STRACHULSKI, J.; FLORIANI, N. 2013. Popular knowledge about plants: an ethnobotanical study in the rural community of linha criciumal, in cândido de abreu-Pr. Revista Geografar. 8(1): 125-153.

SUR, R. .; MARTIN, K. .; LIEBEL, F. .; LYTE, P. .; SHAPIRO, S. .;SOUTHALL, M. 2009. Anti-inflammatory activity of parthenolide-depleted Feverfew (Tanacetum parthenium). Inflammopharmacology. 17(1): 42-9.

SYLVESTRE, S. L.; ROSA, M. M. T. 2002. Methodological manual for botanical studies in the Atlantic Forest. Seropédica: Federal Rural University of Rio de Janeiro. pp. 7-8.

SZABÓ T, A. 1996. Human diversity and plant genetic diversity in the evolution of crop plants. Ethnobiodiversity. In: R. FRITSCH; K. HAMMER (Eds.), Evolution und Taxonomie von pflanzengenetischen Ressourcen. Festschr. fur Peter Hanelt. ZADI, Bonn, Schriften zu Genetischen Ressourcen v. 4, pp. 130-161.

TAMURA, E. K.; JIMENEZ, R. S.; WAISMAM, K.; GOBBO-NETO, L.; LOPES, N. P.; MALPEZZI-MARINHO, E, A.; MARINHO, E. A.; FARSKY, S. H. 2009. Inhibitory effects

of Solidago chilensis Meyen hydroalcoholic extract on acute inflammation. Journal of Ethnopharmacology. v. 122, pp. 478-485.

TAHER, Y.A. 2012. Antinociceptive activity of Mentha piperita leaf aqueous extract in mice Libyan Journal of Medicine. 7 (1): 1-5.

TANG, L. I. C. ; LING, A. P.K.; KOH, R. Y.; CHYE, S. M.; VOON, K. GL. 2012 . Screening of anti-dengue activity in methanolic extracts of medicinal plants. International Society for Complementary Medicine Research. 12(3): 1- 10.

TAVEIRA, L. R. S. 2010. AGRONOMIC INSPECTION REPORT FOR THE FORTALEZA DE SANT'ANA FARM. Ministry of Agrarian Development. National Institute for Colonisation and Agrarian Reform. Regional Superintendence of Minas Gerais. Land Acquisition Division. Belo Horizonte, MG.

TEIXEIRA, M, T. 2012. Olga Benário Settlement: A case study of the spatialisation of the struggle for land in the Zona da Mata region of Minas Gerais. Dissertation submitted to the Federal University of Viçosa (Postgraduate Programme in Rural Extension). Federal University of Viçosa.

TICKTIN, T, GANESAN, N. R.; PARAMESHA M,; SETTY, S. 2012. Disentangling the effects of multiple anthropogenic drivers on the decline of two tropical dry forest trees. Journal of Applied Ecology. 49:774-784.

TOLEDO, V.; BARRERA-BASSOLS, N. 2008. La memori BIOCULTURAL: In importância ecológica de las sabidurias tradicionales. Barcelona, Icaria editorial.

TOLEDO, V. M.; BARRERA-BASSOLS, N. Ethnoecology: a post-normal science that studies traditional knowledge. In: SILVA, V. A; ALMEIDA, A. L. S.; ALBUQUERQUE, U. P. (Orgs.) Ethnobiology and Ethnoecology: people and nature in Latin America. Recife - PE: NUPEEA, 2010 (Current events in ethnobiology and ethnoecology series, v. 1).

TOLEDO, V. M. 2002. Ethnoecology: a conceptual framework for the study of indigenous knowledge of nature. Pp. 511-522. In: Ethnobiology and Biocultural Diversity. Georgia, International Society of Ethnobiology.

TOMAZI, L. B.; AGUIAR, P. A.; CITADINI-ZANETTE, V.; ROSSATO, A. E., 2014. Ethnobotanical study of medicinal trees in the José Milanese Municipal Ecological Park, Criciúma, Santa Catarina, Brazil. Revista brasileira de plantas medicinais, 16(2, supl. 1), pp.450-461.

TORRES-AVILEZ, W. 2014. Gender and age. In: ALBUQUERQUE, U. P. (Org.). Introduction to Ethnobiology. Recife - PE: NUPEEA, 20: 163-167.

VARGHESE, A.; T. TICKTIN.; MANDLE, L. 2015. Importance of, and approach for, assessing the effects of multiple stressors on the regeneration of fruit harvested trees in a tropical dry forest. PLOS. 1-19.

VASCONCELOS, C. M.; VASCONCELOS, T. L. C.; PÓVOAS, F. T. X.; SANTOS, F. E. P.; MAYNART, W. H. C.; ALMEIDA, T. G.; OLIVEIRA, J. F. S.; FRANÇA, A. D. D.; VERÍSSIMO, R. S. S.; LINS, T. H.; ARAÚJO-JÚNIOR, J. X.; BASTOS, M. L. A. 2014. Antimicrobial, antioxidant and cytotoxic activity of extracts of Tabebuia impetiginosa (Mart. ex DC.) Standl. Journal of Chemical and Pharmaceutical Research. 6(7): 2673-2681. 2014.

VEIGA JUNIOR, V. F. 2008. Study of the consumption of medicinal plants in the Centre-North Region of the State of Rio de Janeiro: acceptance by health professionals and mode of use by the population. Revista brasileira de farmacognosia. 18(2): 308-313.

VELOSO, C. C.; OLIVEIRA, M.C.; OLIVEIRA, C.C.; RODRIGUES, V, G .; GIUSTI- PAIVA, A .; DUARTE, I.D.; FERREIRA, A,V.; PEREZ, A. C . 2014. Hydroethanolic extract of Pyrostegia venusta (Ker Gawl.). Miers flowers improves inflammatory and metabolic dysfunction induced by high-refined carbohydrate diet. Journal of Ethnopharmacology 151 (1):722-728.

VENDRUSCOLO, G. S.; MENTZ, L. A. 2006. Ethnobotanical survey of plants used as medicinals by residents of the Ponta Grossa neighbourhood, Porto Alegre, Rio Grande do Sul, Brazil. Iheringia, Série Botânica, 61(1-2), 83-103.

VIEIRA, R. F.; SILVA, S.R. 2002. Strategies for the conservation and management of genetic resources of medicinal and aromatic plants - results of the 1ª technical meeting. EMBRAPA/IBAMA, CNPq, Brasília.

VIU, A.F.M.; VIU, M.A.O.; CAMPOS, L. Z. O.; SANTOS, C. S. 2010. Ethnobotany: a gender issue? Revista Brasileira de Agroecologia, Porto Alegre 5: 138-147.

VOEKS, R. A.; LEONY, A. 2004. Forgetting the Forest: assessing medicinal plant erosion in Eastern Brazil. Economic Botany. 58: 294-306.

WANG, C.; XU, F.Q.; SHANG, J.H.; XIAO, H.; FAN, W.W.; DONG, F.W.; HU, J.M.; ZHOU J. 2013. Cycloartane triterpenoid saponins from water soluble of Passiflora edulis Sims and their antidepressant-like effects. Journal of Ethnopharmacology. 148 (3): 812-817.

YUAN, L. F.; CHEN, L.; LU, P. ; DOU, H.; BO, M.; ZONG, M,M. ; XIA, L, J. 2008. Protective effects of total flavonoids of Bidens pilosa L. (TFB) on animal liver injury and liver fibrosis. Journal of Ethnopharmacology. 116:539-546.

YUAN-YUAN DENG.; YANG YI.; LI-FANG ZHANG.; RUI-FEN ZHANG.; YAN ZHANG.; ZHEN-CHENG, WEI.; XIAO-JUN TANG.; MING-WEI ZHANG. 2014. Immunomodulatory activity and partial characterisation of polysaccharides from Momordica charantia. Molecules. 19 (9): 13432-13447.

ZERAATI, F.; ASHARI, F.E.; ARAGHCHIAN, M.; EMAM, A.H.; RAD, M.V.; SEIF, S.; RAZAGHI, K. 2014. Evaluation of topical antinociceptive effect of Artemisia absinthium extract in mice and possible mechanisms. Afr J Pharm Pharmacol. 8:492-6.

ZETOLA, M.; LIMA, T. C. M.; SONAGLIO, D.; GONZALEZ-ORTEGA, G.; LIMBERGER, R. P.; PETROVICK, P. R.; BASSANI, P. 7 R. 2002. CNS activities of liquid and spray-dried extracts from Lippia alba-Verbenaceae (Brazilian false melissa). Journal of Ethnopharmacology 82, 207-215.

ANNEXES

Annex 1: **Approval of the project by the Research Ethics Committee of the Federal University of Juiz de Fora.**

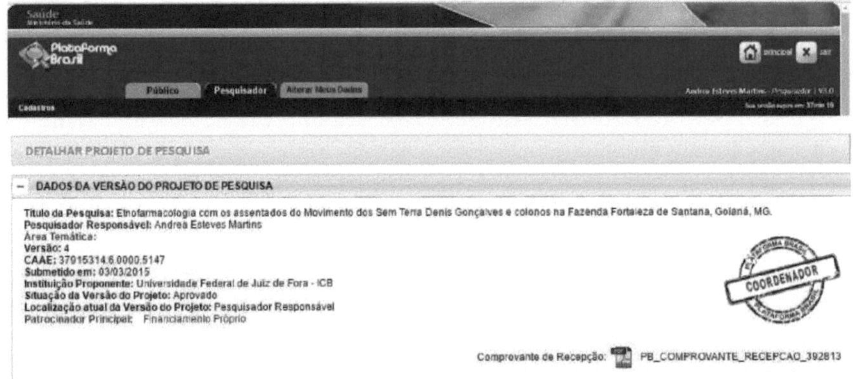

DETALHAR PROJETO DE PESQUISA

– DADOS DA VERSÃO DO PROJETO DE PESQUISA

Título da Pesquisa: Etnofarmacologia com os assentados do Movimento dos Sem Terra Denis Gonçalves e colonos na Fazenda Fortaleza de Santana, Goianá, MG.
Pesquisador Responsável: Andrea Esteves Martins
Área Temática:
Versão: 4
CAAE: 37915314.6.0000.5147
Submetido em: 03/03/2015
Instituição Proponente: Universidade Federal de Juiz de Fora - ICB
Situação da Versão do Projeto: Aprovado
Localização atual da Versão do Projeto: Pesquisador Responsável
Patrocinador Principal: Financiamento Próprio

Comprovante de Recepção: PB_COMPROVANTE_RECEPCAO_392813

Annex 2: Semi-structured form on the use of medicinal plants in the Denis Gonçalves Settlement, Goianá, Minas Gerais.

ETHNOPHARMACOLOGICAL FORM
1-Name of interviewee 2-Date of birth: 3-Occupation: 4-Sex 5-What is your religion? 6-0Origin (where did you come from)? 7-Who did you learn about medicinal plants from? ( ) parents ( )avos ( ) Uncles ( ) other relatives ( )benzedores ( ) root vegetables,

( ) midwives
( ) Other, who
( ) books
( ) magazines
( ) reports
( )radio
( ) television
( ) courses
8-Age group and sex of the people who fall ill the most in the settlement

	Child	Adult	Elderly	
Man				
Woman				

9-What health problems are most common in the settlement?
10-Who is learning about plants with you?
11-Besides plants, do you use any animals or minerals as medicine?
12-Do you use industrialised medicines?
13- In your opinion, which is the better remedy, medicinal plants or industrialised medicines? Why is that?
14-Do you have access to the public health system?
15- Do people in the settlement use the formal public health system?
16-Are your family members continuing your knowledge of medicinal plants?
17-Do you believe that a garden with medicinal plants could bring benefits to the community?
() yes () no
18-Do you believe that ancient knowledge about the use of medicinal plants is valid and that it really works?
() yes () no
19- Do you think it's important to carry out work on nature conservation in the settlement? Would you like to take part? () yes ( ) no () I've never thought about it ( ) I've never heard anything about it.
20-Do you think it's important to hold meetings to exchange knowledge about medicinal plants in the settlement?

Printed by Books on Demand GmbH, Norderstedt / Germany